Topology-Based Modeling of Textile Structures and Their Joint Assemblies

Yordan Kyosev

Topology-Based Modeling of Textile Structures and Their Joint Assemblies

Principles, Algorithms and Limitations

Springer

Yordan Kyosev
Faculty of Textile and Clothing Technology
Hochschule Niederrhein, University
of Applied Sciences
Mönchengladbach, Germany

ISBN 978-3-030-02540-3 ISBN 978-3-030-02541-0 (eBook)
https://doi.org/10.1007/978-3-030-02541-0

Library of Congress Control Number: 2018960466

This Springer imprint is published by the registered company Springer Nature Switzerland AG
The registered company address is: Gewerbestrasse 11, 6330 Cham, Switzerland

Preface

This work presents the experience of the author in the area of modelling of textile structures. For the purpose of the habilitation process, it was possible to submit the single papers and a short summary about the relations between these. I preferred to rewrite it and submit as a monograph. This way, all works combined are presented more systematically and the next generation of modelling experts can find the "dos" and "don'ts" in the area in one source, or can at least shorten the trials and avoid the errors in the development of scientific methods.

At the beginning of my Ph.D. time, in 1997, I was sure, it would be easy to create a virtual yarn, collecting masses and springs, or using beam elements with FEM. Throughout the long development process, it became clear that the principle is correct and that it works. Although the principle was clear and working, the practical implementations for the different kinds of structures caused difficulties. These difficulties were caused by the complexity of the modern way of computer assisted investigation of textiles—these require knowledge in several areas in order to built a proper computer model that would, in the end, actually run and become useful for engineers. The main required areas are:

- understanding the textiles and textile processes, as these are the main objects of this investigation;
- good programming skills, because without complex object oriented models, the levels of the interaction cannot be represented well enough;
- good numerical mathematics skills, because the systems of linear or differential equations have to be solved in order to get result;
- good (nonlinear) mechanics knowledge, because the yarns and fibers behave as continuum (fibers) and structures (yarns and textile fabrics);
- tools and knowledge of 3D computer graphics, because without nice visualization, the results cannot be understood and interpreted properly;
- and a bottle of beer (or apple juice, my friends know that I do not like beer) and a place for walking, in order to overcome the frustration from the debugging errors during the development.

This work tries to give a summary on these areas for the modelling of textile structures at four scales or levels—filaments, yarns, fabrics and assembly. Dominating is the scale of the yarn, because the yarn topology determines the main properties of the fabrics. All equations presented in the following text exist in at least three versions—on paper, in at least one programming language like Matlab or C++ and in LaTeX for this work. If during the transfer from one to another version some error occur, the author asks to excuse this and does not give any guarantee for correct results, when using the equations. The author is thankful to everybody for checking the validity of the equations and giving feedback or advice on discovered mistakes.

Mönchengladbach, Germany Yordan Kyosev

Formal Remark

This work is approved from the Faculty of Mechanical Engineering at Chemnitz University of Technology as habilitation work.
Habilitation committee:

Prof. Dr.-Ing. habil. Sophie Gröger, Technische Universität Chemnitz, Chair
Prof. Dr.-Ing. Holger Cebulla, Technische Universität Chemnitz, Reviewer
Prof. Dr.-Ing. Steffen Marburg, Technische Universität München, Reviewer
Prof. D.Sc. Stepan Lomov, Katholieke Universiteit Leuven, Reviewer
Prof. Dr.-Ing. Maik Berger, Technische Universität Chemnitz, Member
Prof. Dr.-Ing. habil. Michael Groß, Technische Universität Chemnitz, Member
Submitted in September 2017, public defence on 13, August 2018.

Acknowledgements

The author would like to express his gratitude to all people, involved in different ways in the development of this work:

- Prof. Dr.-Ing. Holger Cebulla, who motivated me to collect, summarize and submit my works in the form of a habilitation work
- The three reviewers of this works for their willingness, patience and any critical comments
- All members of the faculty of mechanical engineering (Maschinenbau) of the Technical University Chemnitz, for the discussions and their valuable time
- Prof. Dr. Sc. Stepan Lomov, for all valuable discussions and moral support during the years
- Dipl.-Ing.(TH) Wilfried Renkens, for his trust in me, giving me the task to develop the algorithms for warp knitted structures in 2006–2009 and for a very cooperative and the fruitful collaboration at a later time
- Carla Einhaus for making my English text more clear and understandable
- My students and some of them who then became assistants—Katalin Küster, Anna Rathjens, Alena Cordes, Marcel Beiss, Matthias Aurich, Nora Brinkert, for their willingness to work in the areas of the textile structures and helping me with their practical measurements and comparisons to verify my models in the frame of their theses.
- My family, for giving me never ending time slots in which papa sat in front of the computer, answering "just a minute".

Contents

Part I
Introduction

Chapter 1
Introduction and Problem Definition

1.1 Modelling of Textiles

Textiles were used by humans since thousands of years for creating cloths as comfortable environments for their human bodies. Braided ropes made from flax fibers, dated to the year 6300–6150 B.C. [7] have been found, which indicates that the usage of textile products for other applications than clothing goes way back in time. Contrary to the predictions of my chemistry teacher, who said that after years the cloths will be created as a thin polymer layer on the human body by spraying the polymers through a nozzle and no weaving, knitting and sewing will be needed, the textiles—as fiber based structures—established their position in the daily life and still enter new markets and areas. Today, the fiber based structures are used for building replacements of human body parts, for composites in the car and airplane industry, for concrete reinforcements and other areas [1], where in the older time no fibers were applied. The growing applications of fiber based structures also require the development of methods and tools for their engineering design. Several methods were developed during last 100 years and allow the prediction of various properties of some structures, like weight, porosity and maximal density with several approximations and limitations.

Most commonly used and investigated are the methods for the design of woven (Fig. 1.1a) structures. Their interlacement can be investigated based on projections of the horizontal (weft) and vertical (warp) yarns in separated planes. This made the creation of the geometrical models already possible in early years and the first models were published long before computers were introduced [15, 34]. Between 1970 and 1980, these models became even more complex, because the mechanical equilibrium of the yarns with even more properties and details [36] were considered. Despite the knowledge of the principles and the methods for modelling woven structures, at the current time only one software package—Wisetex, is known to be able to predict their mechanical behaviour and to create a realistic geometry of a structure, taking into account the yarn mechanical properties [27, 43]. The professional CAD systems are connected to the machine control and are able to render very good photo realistic

Y. Kyosev, *Topology-Based Modeling of Textile Structures and Their Joint Assemblies*, https://doi.org/10.1007/978-3-030-02541-0_1

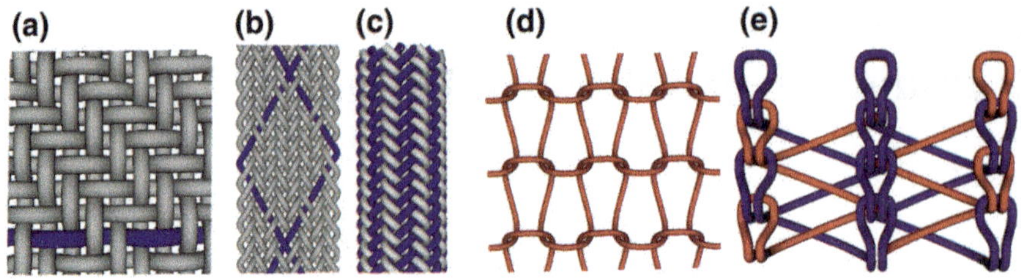

Fig. 1.1 Basic textile (yarn based) structures **a** woven, **b** flat braided, **c** tubular braided, **d** weft knitted, **e** warp knitted

images of the structures. Still, they only cover a small area of fabric structures and do not include any details of the yarn and fabrics mechanics until now.

The braided structures (Fig. 1.1b, c) have the same unit cells as the woven ones. Because of this, there are several works reporting about the unit cell of the braided structures, and only few works, in which the structures are generated using a simulation of the production process [2, 16, 35, 39]. The usage of the simulation process for quick product development is not efficient and the unit cells are not enough for larger structures. The only industrially applicable software implementation for modelling of braided structures, currently available online, was developed by the author [22] and is distributed as TexMind Braider [21].

The knitted structures—both weft and warp knitted—(Fig. 1.1d, e) are built by interlooping of yarns. Since the loops have a complex 3D geometry, their description using projections is more complex and the research works are significantly less, than for woven structures. The basic principle for modelling—based on the few key points, is also described in the pre-computer time [31] and extended by mechanics later [37, 38]. The weft knitted structures consist of one yarn per row (Fig. 1.1d) and computational models and implementations in software can as well be found for more complex structures, having tucks, missing stitches and transfer [14, 29, 30, 42]. All these works usually consider a topological model and do not perform any adjustment of the model, based on the real yarn length in the fabrics. The warp knitted structures (Fig. 1.1d) are produced by interloopings of several yarn groups. Because the complexity of the possible structures is significantly larger, results of only a few groups are reported in this area [13, 25, 41] where implementation in industrial software with proven coverage of really large sets of structures is only known for the models of the team of the author (Kyosev) and Renkens [18–20, 40]. These models are initially implemented in the product Warp3D by ALC Computertechnik, Aachen Germany. Since the year 2009, the software is developed and maintained from the partial successor of ALC—Texion GmbH, Aachen, Germany for which no public information about the methods is found anymore. A revised and extended version of these methods is implemented in the software of TexMind UG [23], developed by the author.

As this short overview demonstrates, the generalized algorithms, methods and software for 3D representation of braided and warp knitted structures, these are mainly developed and completely by the author. This work presents systematical representation of these methods and algorithms with their advantages and limitations.

1.2 Strategies for Multiscale Modelling of Textiles

The modelling of the textiles multiple scales—for instance yarn and fiber/filament level, requires separate algorithms and methods for each scale. The knitted, woven and braided structures are yarn based and require certain descriptions of the yarn geometry. There are three popular methods for retrieving the yarn geometry:

- By using image processing of 2D or 3D images, for instance by X-ray micro-computer tomography [10, 32]. This method can produce accurate data about the geometry of the fibrous structure and the orientation of the fibers, but is only applicable for already produced structures. Due to this limitation, it will not be further discussed within this work.
- Simulating the complete production process [2, 6]. This method is currently still connected with many computations and with a lot of simulation time. Nevertheless, considering the rapid development of the computer technic and methods for parallel processing, it is going to become more popular in the near future.
- Generating the topology of the structure parametrically and refining it by using some mechanical methods. This approach is the most common one and allows the generation of a textile geometry in a very short time. One of the first works regarding woven structures are the ones of Peirce and Kemp [15, 34]. Several geometrical methods are summarized by Behera and Hari in [3]. The modern methods considering the mechanical properties of the yarns are usually based on the minimization of the potential energy of the yarns, as described in the book of Postle et al. [36]. In the software Wisetex [26–28], the mechanical models for woven and braided structures are implemented. The geometry of the loops in knitted structures is investigated by researchers as for instance Leaf [24], Postle and Munden [37, 38], Hart et al. [11, 12], Goktepe and Harlock [8], Wu [44]. A software, useful for 3D modelling of knitted structures, is the Weft Knit (part of Wisetex) by Moesen [30] and the different computer realizations of Kyosev and Renkens [17, 18, 40]. Very good photorealistic simulations of knitted fabrics at yarn level are reported by Kaldor et al. [14]. They are extended for complete clothing by Yuksel et al. [45].

Both the techniques, being able to create virtual products without using existing ones—the simulation of the complete production process and generating topology of the structures—normally require the use of the yarns as a main object at the first step. As a second step, the created yarns are filled with filaments and eventually refined by consideration of the contact between the filaments. The distribution of the fibers or filaments in the yarns can be based on statistical distribution or based on some of the methods used by Neckar and Das [33], Grishanov and Lomov [9], using complete FEM calculations in which the fibers are represented as beams as done by

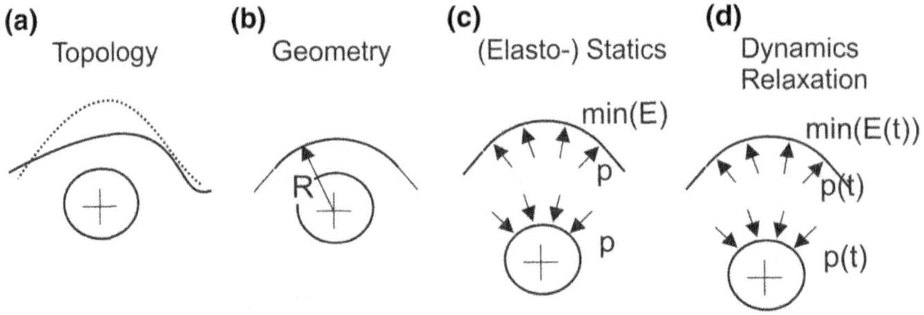

Fig. 1.2 Different types of models for description of the geometry of textile structures

Durville in his Multifil package [4, 5] or arbitrary arrangement of the single filaments in circular or parallel layers as implemented in the packages of the TexMind software and described by Kyosev in [22].

This work, concentrates on the topological methods. Here the meaning of "topology" is the knowledge of the orientation and positions of the yarns (or their axes), related to the other yarns in the same structure. The topology of the mathematical meaning, including knot theory etc. does not really help at the current state in the generation of textile structures. This can turn into the correct scientific approach, but its methods have to make further developments and reach an applicable level. The methods used here are named "topological" and not "geometrical", because they do not pretend to give *exact* geometric description of the position of the yarns. The topological methods define that one yarn (curve) in Fig. 1.2a crosses another (presented with its cross section, as circle), but do not define explicitly that the curve is an circular arc, as the case in Fig. 1.2b. The geometrical methods would be more accurate in the description, but they will be automatically more limited as well. The circular arc of the case in Fig. 1.2b is only valid for crossing yarns with circular cross section under some tension. If the yarns are deformable (and in the most cases these are deformable) the curve changes to another type. Because this is a common case, for the purpose of this work and for the purpose of the creation of industrial CAD systems with general use, it was assumed, that the geometrical methods would lead to more limitations and complexity. The real axial curve of the yarns depends of course on the mechanical properties of the yarns, the stresses in these and on the load history of the structures. For calculations of these computational mechanics methods are required, based on force equilibrium or minimum of the potential energy of the system (Fig. 1.2c). For the long term behavior, the factor time has to be included and additionally the dynamical effects for high speed loadings and the relaxation processes, including damages, have to be considered (Fig. 1.2d). This book limits the content to creating the topology based description of the structures only. Some methods for applications of the mechanical models are demonstrated in the last chapters.

1.3 Content of the Book

The book has seven parts, of which the first one contains this introduction. Part II covers the braided structures, presenting generalized models for their topology and application of this model for evaluation of their properties. A separated chapter deals with universal topological approach for custom braiding machines, in which the topological orientation is created based on the machine emulation. This method is more computational intensive, but allows the generation of the geometry for braiding machines with complex configuration.

Part III is dedicated to the knitted structures. Its main chapter covers the topological construction of the structural elements of the warp and weft knitted structures. The next chapter presents one method for the computation of knitted structures, based on a mesh or truss framework with its implementation issues, advantages and problems.

Part IV gives an overview about the modelling of sewing stitches, as an important element for the connection of fabrics into assembly. In the same part a small chapter about woven structures is hosted, only for the sake of completeness. It consists of a few references and explanations on why the woven structures were not a primary goal of investigation of the author.

Part V consists of the extension of the yarn level to the filament, fabrics and assembly. The first chapter presents implementation issues about the multifilament modelling and rendering of yarn structures. The second chapter is dedicated to the placing of structural cells in the space and orientation of the fabrics before the sewing yarns are added. The last chapter in this part gives an overview about the implementations of the algorithms in software—selection of the environment, levels in the structures and comparative analysis to another packages.

Part VI "Mechanics after topology" discusses the application of mechanical models for refinement of the topologically generated textile structures, implemented by the author. The last chapter presents applications of geometrical models, based on external computational tools like FEM or digital chain software.

References

1. Adanur, S.: Wellington Sears Handbook of Industrial Textiles. Technomic Publishing, Lancaster, PA (1995)
2. Akkerman, R., Villa, R.B.H.: Braiding simulation for RTM preforms. In: Long, A. (ed.) TexComp8: 8th International Conference on Textile Composites, Nottingham, UK (2006)
3. Behera, B., Hari, B.K.: Woven Textile Structure: Theory and Application. Woodhead Publishing Ltd, Cambridge (2010). http://www.worldcat.org/oclc/610819829
4. Durville, D.: Finite element simulation of textile materials at mesoscopic scale. In: Finite Element Modelling of Textiles and Textile Composites, St-Petersburg, 26–28 Sept 2007
5. Durville, D.: Simulation of the mechanical behaviour of woven fabrics at the scale of fibers. Int. J. Mater. Form. **3**(S2), 1241–1251 (2010). https://doi.org/10.1007/s12289-009-0674-7
6. Finckh, H.: Textile micromodels as a result of idealized simulation of production processes. In: Finite Element Modeling of Textiles and Textile Composites. Katholieke Universiteit Leuven & St-Petersburg State University of Technology and Design, Saint-Petersburg and Russia (2007)

7. Gerichhausen, H.: Das blaue Wunder der Region Heinsberg, 1, aufl edn. Heimatverein Wegberg-Beeck, Wegberg (2011)
8. Goktepe, O., Harlock, S.: Three-dimensional computer modeling of warp knitted structures. Text. Res. J. **72**, 266–272 (2002)
9. Grishanov, S., Lomov, S.E.A.: The simulation of the geometry of two-component yarns. Part I: The mechanics of strange compression: simulating yarn cross-section shape. J. Text. Inst. **88**, Part 1 (2) (1997)
10. Harjkova, G., Barburski, M., Lomov, S.V., Kononova, O., Verpoest, I.: Weft knitted loop geometry of glass and steel fiber fabrics measured with X-ray micro-computer tomography. Text. Res. J. **84**(5), 500–512 (2013). https://doi.org/10.1177/0040517513503730
11. Hart, K., de Jong, S., Postle, R.: Analysis of the single bar warp knitted structure using an energy minimization technique: Part I: Theoretical development. Text. Res. J. **55**(8), 489–498 (1985). https://doi.org/10.1177/004051758505500807. http://trj.sagepub.com/cgi/content/abstract/55/8/489
12. Hart, K., de Jong, S., Postle, R.: Analysis of the single bar warp knitted structure using an energy minimization technique: Part II: Results and comparison with woven and weft knitted analysis. Text. Res. J. **55**(9), 530–539 (1985). https://doi.org/10.1177/004051758505500903. http://trj.sagepub.com/cgi/content/abstract/55/9/530
13. Honglian, C., Mingqiao, G., Gaoming, J.: Three-dimensional simulation of warp-knitted fabric. Fibres Text. Eastern Eur. **17**(3 (74)), 66–69 (2009)
14. Kaldor, J., James, D.L., Marschner, S.: Simulating knitted cloth at the yarn level. In: Proceedings of SIGGRAPH 2008. Held in Los Angeles, California, Aug 2008 (2008)
15. Kemp, A.: An extension of peirce's cloth geometry to the treatment of non-circular threads. J. Text. Inst. Trans. **49**(1), T44–T48 (1958). https://doi.org/10.1080/19447025808660119. http://www.tandfonline.com/doi/abs/10.1080/19447025808660119
16. Kessels, J., Akkerman, R.: Prediction of the yarn trajectories on complex braided preforms. Compos. Part A: Appl. Sci. Manuf. **33**(8), 1073–1081 (2002). https://doi.org/10.1016/S1359-835X(02)00075-1
17. Kyosev, Y.: Computational model of loops of a weft knitted fabric. In: Dragčević, Z. (ed.) Magic World of Textiles, Zagreb (2006)
18. Kyosev, Y., Renkens, W.: 3D-CAD für die gestaltung von gewirkten strukturen. In: 11-Chemnitzer Textiltagung, 24–25 Oktober 2007, pp. 110–117 (2007)
19. Kyosev, Y., Renkens, W.: Prediction of some properties of warp knitted structures based on numerical simulation. In: Dragcevic, Z. (ed.) Book of Proceedings for the 4th International Textile, Clothing and Design Conference Magic World of Textiles (2008)
20. Kyosev, Y., Renkens, W.: Modelling and visualization of knitted fabrics. In: Chen, X. (ed.) Modelling and Uredicting Textile Behaviour, pp. 225–262. Woodhead Publishing and In association with the Textile Institute and CRC Press, Cambridge and Boca Raton and FL (2010)
21. Kyosev, Y.K.: Texmind Braider, (2012). www.texmind.com
22. Kyosev, Y.K.: Generalized geometric modelling of tubular and flat braided structures with arbitrary floating length and multiple filaments. Text. Res. J. **86**(12), 1270–1279 (2015)
23. Kyosev, Y.K.: Texmind Warp Knitting Editor 3D (2017). www.texmind.com
24. Leaf, G.G.A.: The geometry of a plain knitted loop. J. Text. Inst. **45**, T587–605 (1955)
25. Li, X., Jiang, G., Ma, P.: Computer-aided design method of warp-knitted jacquard spacer fabrics. Autex Res. J. **16**(2) (2015). https://doi.org/10.1515/aut-2015-0027
26. Lomov, S.: Integrated textile preprocessor WiseTex, version 3.2. In: Computational Models, Methods and Algorithms. KU Leuven, Leuven (2013)
27. Lomov, S., Bernal, E., Koissin, V., Peeters, T.: Integrated textile preprocessor wisetex, version 2.5. In: Computational Models, Methods and Algorithms
28. Lomov, S., et al.: Wisetex (2011). http://www.mtm.kuleuven.be/Onderzoek/Composites/Research/meso-macro/textile_composites_map/textile_modelling/textile_modelling_fe
29. Meißner, M., Eberhardt, B.: The art of knitted fabrics, realistic and physically based modelling of knitted patterns. In: EUROGRAPHICS '98, vol. 17, no. 3 (1998)

30. Moesen, M., Lomov, S., Verpoest, I.: Modelling of the geometry of weft-knit fabrics. In: Techtextil (Hg.) 2003—TechTextil Symposium, 7–10 Apr 2003 (2003)
31. Munden, D.: The geometry and dimensional properties of plain-knit fabric. J. Text. Inst. **50**, T448–T471 (1959)
32. Naouar, N., Vidal-Sallé, E., Schneider, J., Maire, E., Boisse, P.: Meso-scale FE analyses of textile composite reinforcement deformation based on X-ray computed tomography. Compos. Struct. **116**, 165–176 (2014). https://doi.org/10.1016/j.compstruct.2014.04.026
33. Neckář, B., Das, D.: Theory of Structure and Mechanics of Fibrous Assemblies. Woodhead Publishing India in Textiles. Woodhead Publishing India, New Delhi (2012)
34. Peirce, F.T.: 5—the geometry of cloth structure. J. Text. Inst. Trans. **28**(3), T45–T96 (1937). https://doi.org/10.1080/19447023708658809. http://www.tandfonline.com/doi/abs/10.1080/19447023708658809
35. Pickett, A.K., Sirtautas, J., Erber, A.: Braiding simulation and prediction of mechanical properties. Appl. Compos. Mater. **16**(6), 345–364 (2009). https://doi.org/10.1007/s10443-009-9102-x
36. Postle, R., Carnaby, G., de Jong, S.: The Mechanics of Wool Structures. Ellis Horwood Limited (1988)
37. Postle, R., Munden, D.: Analysis of the dry-relaxed knitted-loop configuration Part 1: two-dimensional analysis. J. Text. Inst. **58**(8), 329–351 (1967)
38. Postle, R., Munden, D.: Analysis of the dry-relaxed knitted-loop configuration Part 2: three-dimensional analyses. J. Text. Inst. **58**(8), 352–365 (1967)
39. Ravenhorst, J.H. van, Akkerman, R.: A spool pattern tool for circular braiding. In: 18th International Conference on Composite Materials, ICCM 2011. Jeju Island, Korea (2011). http://doc.utwente.nl/78415/
40. Renkens, W., Kyosev, Y.: Geometrical modelling of warp knitted fabrics. In: Finite Element Modelling of Textiles and Textile Composites, St-Petersburg, 26–28 Sept 2007, CD-ROM Proceedings (2007)
41. Robitaille, F., Clayton, B.R., Long, A.C., Souter, B.J., Rudd, C.D.: Geometric modelling of industrial preforms: warp-knitted and multiple layer textiles. J. Mater. Des. Appl. Proc. Inst. Mech. Eng. (Part L) **214**, 71–90 (2000)
42. Shima Seiki: SDS-ONE APEX (2011)
43. Verpoest, I., Lomov, S.V.: Virtual textile composites software wisetex: integration with micro-mechanical, permeability and structural analysis. Compos. Sci. Technol. **65**(15–16), 2563–2574 (2005)
44. Wu, W., Hamada, H., Maekawa, Z.: Computer simulation of the deformation of weft knitted fabtrics for composite materials. J. Text. Inst. **85**(2), 198–214 (1994)
45. Yuksel, C., Kaldor, J.M., James, D.L., Marschner, S.: Stitch meshes for modeling knitted clothing with yarn-level detail. ACM Trans. Graph. **31**(4), 1–12 (2012). https://doi.org/10.1145/2185520.2185533

Part II
Braided Structures

Chapter 2
Topology Based Models of Tubular and Flat Braided Structures

2.1 Introduction

This chapter presents topology based models for the most used braided structures—tubular and flat braids. The machine types and the rules for their configuration, including the horn gears and carriers, are well described and analysed and it is shown, that the geometry of the braids can be predicted by using parametric models. The presented algorithms and modifications significantly exceed the state of the art. They present a generalized approach for all kinds of interlacements of tubular and flat braided structures and allow its implementation in a software for an efficient design of braids.

2.2 State of the Art

In this chapter, only the work related to the macro-scale 3D geometry of braided structures will be discussed. The models, presenting only unit cells of the braided (and woven) fabrics, can be significantly more accurate regarding the unit cell level, but might be incorrect regarding the macro scale level. The reason for this is, that not all (woven) unit cell configurations can be produced on braiding machines and therefore there can be no unit cells of braided structures. During the production of the braided structures, the carriers can not choose their interlacement points arbitrarily, as this is done by the dobby or jacquard machines during the weaving process. The interlacement depends on the horn gear configuration, the track and the carrier arrangement. The **natural** way for preparing the geometry (and unit cell) of braided structures require consideration of these three groups of information. The consideration of the exact machine configuration is an important advantage of the developed method.

The modelling of textiles in 3D using computer systems, is described by Liao and Adanur [13]. The idea is to sweep a simple 2D closed contour $c(u)u \in [0, M]$ along

Y. Kyosev, *Topology-Based Modeling of Textile Structures and Their Joint Assemblies*, https://doi.org/10.1007/978-3-030-02541-0_2

a regular 3D curve in the 3D space $\gamma(s),\ s \in [0, L]$ with non-vanishing curvature, in order to receive a tubular surface as sweep object.

$$G(s, u) = \gamma(s) + c_1(u) \cdot n(s) + c_2(u) \cdot b(s) \tag{2.1}$$

where $n(s)$ and $b(s)$ denote the normal and binormal vectors of $\gamma(s)$ respectively. They suggest a modified Frenet frame so that the sweep object is valid for very general cases, which are able to handle trajectory γ consisting of planar and nonplanar segments, as long as the curvature of each nonplanar segment is not zero. This method in its digitalized form (where the normals and binormals are computed based on discrete data set) presents the basic principle of the modelling of yarn surfaces of textile structures. Implementing this method in C++ and using OpenGL, the author created a realistic looking 3D simulations of braided fabrics with different structures around conical and other pulled preforms of diamond and regular braids. How the coordinates of the yarns are computed exactely, is not given in the paper.

The derivation of the equations for the yarn paths of open braids is given by Rawal et al. [20, 21], in which the undulations of the yarn is not considered. For instance for a cylinder these are in the form.

$$\begin{aligned} x &= r \cdot \cos(\theta) \\ y &= \pm r \cdot \sin(\theta) \\ z &= r \cdot \theta \cdot \cot(\alpha) \end{aligned} \tag{2.2}$$

where r is the nominal radius of the braid and $\theta = \omega \cdot t$ is the angle of the investigated point, depending on the angular velocity of the carrier ω around the product axis.

Kyosev et al. [6, 10] used an extended version of these equations, considering the undulations of the radius r in the places of contact points.

$$r = r_{mandrel}(z) + d_{yarn} \pm 0.5 \cdot d_{yarn} \tag{2.3}$$

The profile of the mandrel is defined as a function in discretised form $r_{mandrel}(z)$ before the calculations of each crossing point of the yarns and the stability conditions of the yarns over the mandrel are checked. The check is based on the equation for winding bodies from Proshkov [18], based on the geodesic angle of the curve τ_G, the limit friction angle ε_{max}, determined from the static friction coefficient between the yarn and mandrel μ and the braiding angle α:

$$\tan\alpha \leq \tan\tau_G \leq \tan\varepsilon_{\max} = \mu \tag{2.4}$$

In this case, the yarns are stable and do not slip around the crossing points, the complete set of yarn paths is generated and connected using splines, in a similar way to those, reported by Pastore et al., who used Bezier curves [17] or Bogdanovich [2, 3] and Lomov [15]. This algorithm is tested for parts with rotational symmetry, for which the stability condition has to be tested once per crossing point with the same

Fig. 2.1 Geometric model of the braid of overbraided mandrel with non constant cross section with wrong oriented ridges. Simulated image from [10]

Z coordinate and for which the geometry of braids using an elliptical cross section of the yarns is created as presented in the Fig. 2.1.

Alpyildiz [1] makes use of a pure analytical description of the curves of the yarn path and describes in detail the equations for the yarn undulations. For regular braids for instance, the regions of the floating are fixed to a high $\pm a/2$ and the region between these is presented with a sinus function, so that the equation for the radius r for the yarns, in which carriers move in counter-clockwise direction looks like:

$$r(\theta) = \begin{cases} \frac{a}{2}; & 0 < \theta < k_1\beta/2 \\ \frac{a}{2} \cdot \sin\left(\frac{2\pi}{\beta}\theta + \frac{\pi}{2}\right); & \frac{k_1\beta}{2} < \theta < (k_1+1)\beta/2 \\ -\frac{a}{2}; & (k_1+1)\beta/2 < \theta < (2k_1+1)\beta/2 \\ \frac{a}{2} \cdot \sin\left(\frac{2\pi}{\beta}\theta + \frac{\pi}{2}\right); & (2k_1+1)\beta/2 < \theta < (2k_1+1)\beta \end{cases} \tag{2.5}$$

and for those running clockwise in similar way, with changed sign of a/2 and changed phase in the sinus function regions [1].

Rawal et al. [19] extends and combines these equations for the cases of using a mandrel of conical cross section and parts, which consist of a combination of cylindrical and conical regions.

The papers cited above, concentrate on the modeling of three main braiding structures diamond, regular and Hercules-braided structures covering biaxial and triaxial braids up to a floating length of three. Unit cells with arbitrary floating length can be created with Wisetex [14, 16], but it is the responsibility of the **user** of Wisetex to obtain the **correct** topology (from the braiding point of view) and the correct orientation of the unit cell in the macro geometry of the braided product. The initial models of Kyosev [6, 7] represent the structures with arbitrary floating length correctly, but they do not represent braids with several yarns in a group realistically. As explained in the mentioned work, these have more space between the yarns, if there are more yarns in a group and thus, can be used for understanding the braid, but not to execute extended calculations.

2.3 3D Models of Braids with Floating Length of Two (Regular Braids)

In the modeling works of several authors [1, 10, 19, 22], a small error regarding the models of regular braids appears. This error is only recognizable by experienced braiders. The mentioned fabrics cannot be produced on any "normal" classical maypole braiding machine with horn gears. The rectangles in Fig. 2.1 point to very well visible ridges that are perpendicular and not parallel to the product axis. Figure 2.2a schematically visualizes once more the orientation of the ridges in the above-mentioned papers. The visible yarn pieces, which build the visible ridges, are placed in square mesh as this is done usually during the drawing of the draft of the structure [4, 5, 7, 12]. In a standard machine, each horn gear rotates in only one direction. All the carriers moving outside of one horn gear are building a vertical ridge, which is running parallel to the take-off speed vector and to the product axis as well (Fig. 2.2b). Therefore, during the analysis of the braids, the ridges are counted and their number gives the number of the horn gears. In order to produce horizontally oriented ridges, the horn gears have to move the carriers in alternating directions, meaning the first carrier moves in one direction and the next one carrier in the opposite direction (Fig. 2.2a). This configuration is only possible on only few in the world 3D braiding machines with individual drives of the horn gears. It is not possible for maypole braiding machines to produce the mentioned models, because all horn gears are connected with certain gears and solely move without changing the rotation direction during the production.

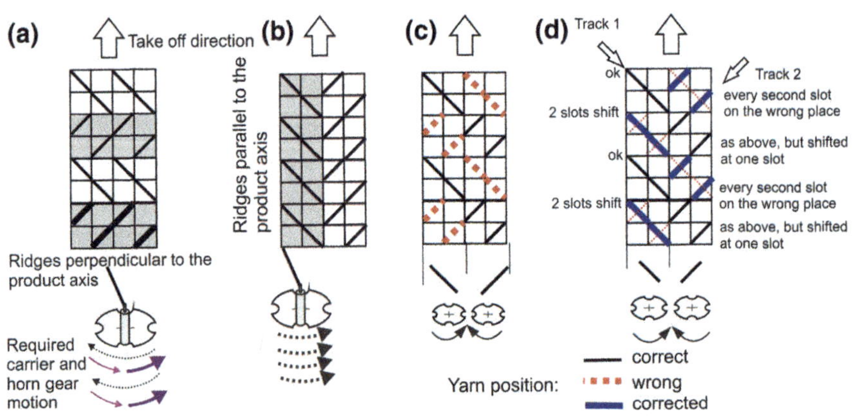

Fig. 2.2 **a** Orientation of the ridges in some models of regular braids, **not producible** on normal braiding machines, **b** proper orientation of the ridges of regular braids, **c** dotted lines show the wrong pieces, **d** image with the wrong and corrected yarn pieces, which can be used for adjusting the starting angle of trigonometrically based models for regular braids. Initial version of the image published in [9], Kyosev, Yordan, Generalized geometric modeling of tubular and flat braided structures with arbitrary floating length and multiple filaments, Textile Research Journal, Vol. 86, issue 12, pages 1270–1279, Copyright (C) 2015 SAGE, Reprinted by permission of SAGE Publications

Topologically, the unit cells of both structures are equivalent, but one structure is simply rotated at 90°. Technologically, a braid with a (extended) unit cell like in Fig. 2.2a is not identical to a braid with a unit cell from case (b) at the macro level. The models of the structures with horizontal ridges cannot be used for color patterning of tubular braids. Furthermore, their use for mechanical calculations can lead to some differences in the mechanical behavior. After analyzing the modeled pictures and comparing the proper yarn positions with the modeled ones, it was possible to detect the positions in which the yarns were placed on the wrong side. These positions are marked and illustrated in Fig. 2.2c with a thick dotted line. It can be seen, that on track 1 (Fig. 2.2d), every second yarn is placed with a phase shifting, which corresponds to the moving of the carrier on two slots around the horn gear. Not counting the yarns in this track consequently, but instead, dividing them into two groups, in which the second has a phase change of angle, corresponding to angle *pi* of the horn gears, might be a possible solution for the correction of the mathematical models for the regular braids. The yarns on track 2 (Fig. 2.2d) are once on the correct side and once on the wrong side of their cells. Also the starting position is alternating, meaning that one yarn is starting in the proper position and one is starting in a wrong position. Writing the correct phases and numerations of the mentioned models is not the goal of this paper, because the approach presented in the next sections is more intuitive and generalized. The aim is to present a common method for the calculation of braids with any floating length per ridge and any number of yarns in a group (= filaments in the yarn). Being independent of the braiding architecture, this method allows its implementation into industrial softwares as it is not limited to one or another type of braid.

It can be seen, that on track 1, (Fig. 2.2c), every second yarn is placed with a phase shifting, which corresponds to the travelling of the carrier on two slots around the horn gear. Not counting the yarns in this track consequently anymore, but in two groups, in which the second has a phase change of angle, corresponding to angle π of the horn gears, can be one possible solution for the correction of the mathematical models for the regular braids.

All of the yarns of track 2, in the same figure, are on the correct side in one cell and on the wrong side in the other cell. Also the situation of the starting cell is alternating—once starting in the proper position, once starting in the wrong position.

Writing the correct phases and numerations of the mentioned models is not worth it and not the goal of this work, because the content presented in the next sections, is more intuitive and generalized. The goal is to present a common method for the calculation of braids with *any* floating length per ridge and *any* number of yarns in a group (= filaments in the yarn), which works and can be used in industrial applications and which is not limited to one or another type of braid.

2.4 Generalized Model for Yarn Path of Braid with Arbitrary Floating Length

For this model, a part of a generalized braid (tubular or flat) with ridges containing different floating lengths (Fig. 2.3) is considered. This part consists of five visible ridges, in which the first, second and the fifth have a floating length of two, while the third ridge has a floating length of five and the fourth has a floating length of one. The sequence of the floating lengths per ridge can be written as 2:2:5:1:2.

The part of the machine that is used for the production of such a braid is presented in the bottom part of Fig. 2.3, following the main braiding equation as a rule for the floating length FL [7]:

$$FL = \left[\frac{Number\_slots\_per\_horn\_gear}{Repeat\_of\_the\_carrier\_arrangement} \right] \tag{2.6}$$

Considering the most common arrangement of 1 full 1 empty, the repeat of the arrangement is 2, so the horn gears should have 4, 4, 10, 2 and 4 slots for this braiding. It should be considered that yarn 1 in Fig. 2.3, floats first (starting from z = 0) under two yarns, then over two, then under five, over one and under two. This

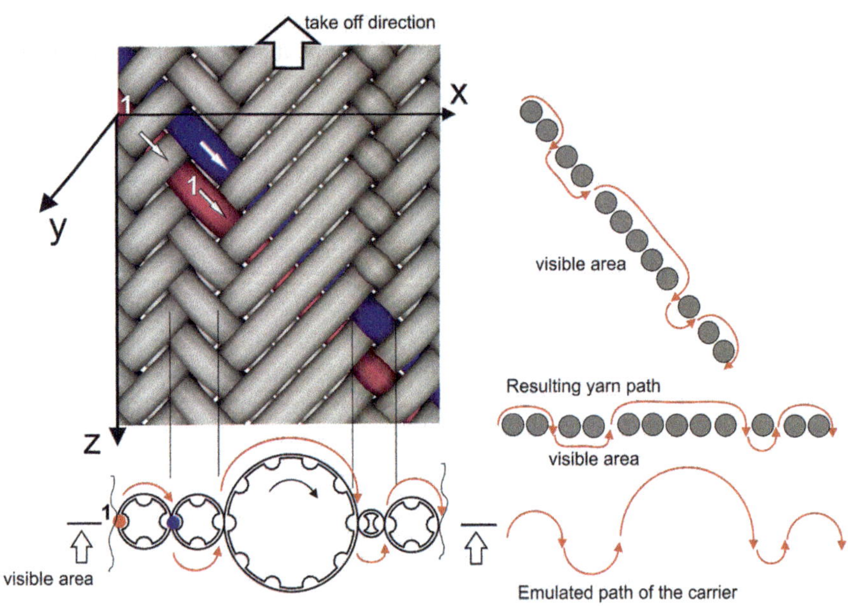

Fig. 2.3 Part of a braid and the corresponding braiding machine for explanation of generalized geometric model of braids. This left image is from [9] Kyosev, Yordan, Generalized geometric modeling of tubular and flat braided structures with arbitrary floating length and multiple filaments, Textile Research Journal, Vol. 86, issue 12, pages 1270–1279, Copyright (C) 2015 SAGE, Reprinted by permission of SAGE Publications

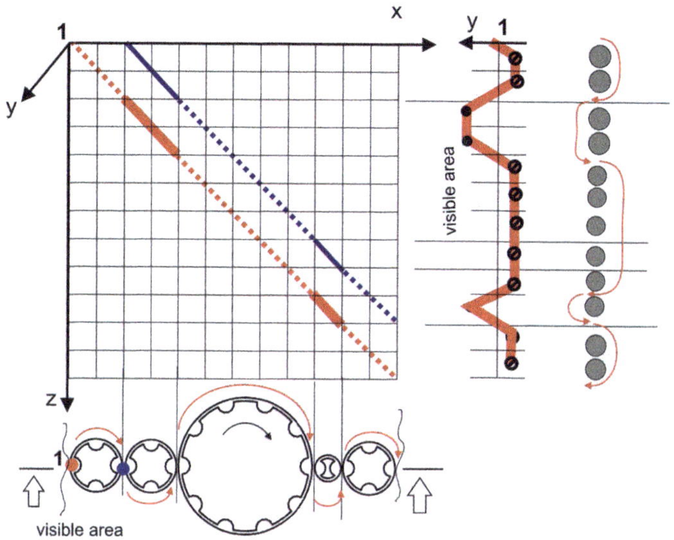

Fig. 2.4 Derivation of the undulation of the yarn coordinate, because of the interlacement with other yarns, demonstrated on the two yarns from Fig. 2.3. The direction Y corresponds to the thickness of the braided product. The left image is from [9]. Kyosev, Yordan, Generalized geometric modeling of tubular and flat braided structures with arbitrary floating length and multiple filaments, Textile Research Journal, Vol. 86, issue 12, pages 1270–1279, Copyright (C) 2015 SAGE, Reprinted by permission of SAGE Publications

floating corresponds to the carrier motion the floating over each yarn corresponds to the carrier motion in the angle, equal to one segment, in case of full occupation of the machine (full occupation means that all possible carrier positions are filled with carriers and correspond to a 1 full 1 empty arrangement along the track [7]. This way, the same number of key points as the number of yarns in the opposite track has to be created for yarn 1, as all these yarns will be crossed. If the current ridge is visible, the key point has a positive Y coordinate; if the ridge is hidden (the yarn is under another) then the key point has a negative Y coordinate. Each ridge with a floating length of two has two key points on the same side, a ridge with a floating length of five has five points and so on. For yarn 1 the first ridge has a floating length of two and it is hidden, so the first two points have a negative Y coordinate; the following ridge is visible and has a floating length of two—so there are two points with positive Y coordinate. Following is a ridge with floating length of five there are five points with a negative Y coordinate; the next ridge is visible and has a floating length of one and the last is hidden and has a floating length of two. In Fig. 2.4 it can be seen that the next yarn starts its motion with an opposite phase. Such specific rules can be derived only for specific floating lengths and not for the general case. An algorithm for the calculation of the coordinates of the key points of a tubular or flat braid will be very similar to the algorithms for a graphical analysis of the braided structures, presented by Engels [4] and Kyosev [7]. This algorithm uses inverse braiding, which means that the horn gears are staying and the carriers are jumping from slot to slot, summarized as follows.

1. Initialize the braiding machine, considering the (minimal) size of the horn gears for the wished floating length of their ridges. The number of slots is stored in a vector $BM_i i \in [1, N_{HG}]$.
2. Initialize the rotation direction of the horn gears $HGR_i i \in [1, N_{HG}]$.
3. Initialize the track(s) of this machine as a list of slots as elements $T(i_{hg}, i_{slot})$, where the carrier slots in one track, denoted with their horn gear number i_{hg} and the slot number i_{slot}, are stored.
4. Fill the tracks with carriers according to the given carrier arrangement, for instance A = [1 0], one full, one empty. The carrier arrangement can be derived from the braiding equation.
5. Determine which slots are currently moving the carriers, which are building visible yarn parts and which are building hidden ones. The connection line between the horn gear centers can be used as the visibility limit.
6. Create the mesh of cells in a manner that provides every two slots of one horn gear corresponding to one column. The mesh size defined by cell height and cell length is easily determinable by using the number of yarns in the braid, the width or the diameter of the braid and the braiding angle [7].
7. For each carrier: go through its track and get the X and Z coordinates of the corresponding cells. If the current slot is visible, set the Y coordinate to +a, if it is not visible to a, and in case of a turning cell (for flat braids) to zero. The variable a can be adjusted depending on the used model for the yarn cross-section and depending on the type of the braid bi- or triaxial. For triaxial braids the yarn configuration of the inlay yarn has to be considered in the variable a.

For flat braids the x coordinate remains on the same plane and for tubular braids it has to be rolled over a cylinder surface, using it as a counter variable for the division of the central angle. Figures 2.5 and 2.6 demonstrate simulated braided structures using this algorithm.

2.5 Multiple Yarns in a Group

The described algorithm presents good results when using single yarns per carrier and a carrier arrangement of one full—one empty of the machine (1F:1E). The yarns are interlacing regularly for this arrangement and the space between each two yarns can be calculated and filled. When one structure is based on more than one yarn in a group, this calculation is not possible anymore with the algorithm, as explained in Sect. 2.4, and as the Fig. 2.7c demonstrates. In Fig. 2.7a is presenting the machine configuration for the production of this braid. It is visible that the carrier arrangement for such a structure is 2 Full–2 Empty. This arrangement is visualized as well in Fig. 2.7b. When this algorithm is applied, the coordinates for the two yarns from the carriers are calculated and the space is reserved correctly, but after that, the space is

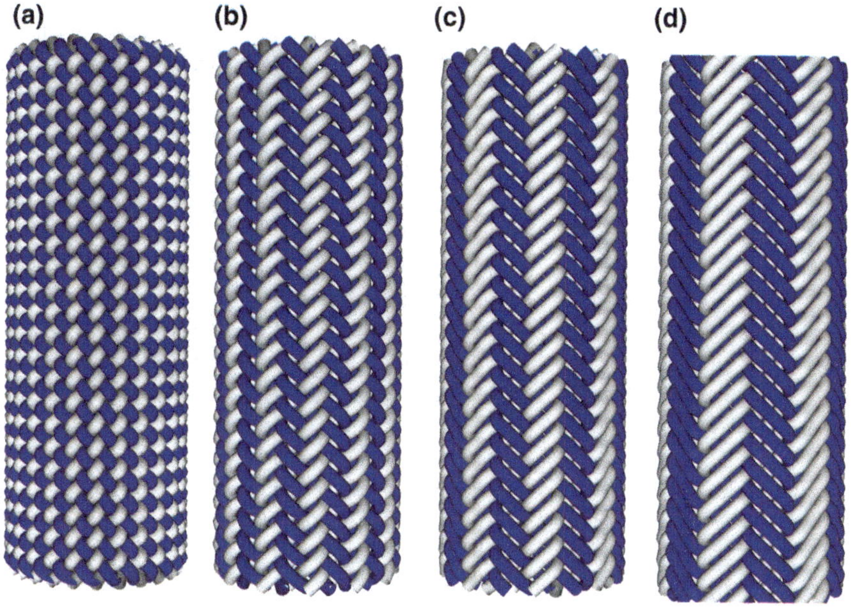

Fig. 2.5 Examples of simulation of tubular braids, all with one yarn in a group, but with different floating length: **a** one, **b** two , **c** three, **d** four

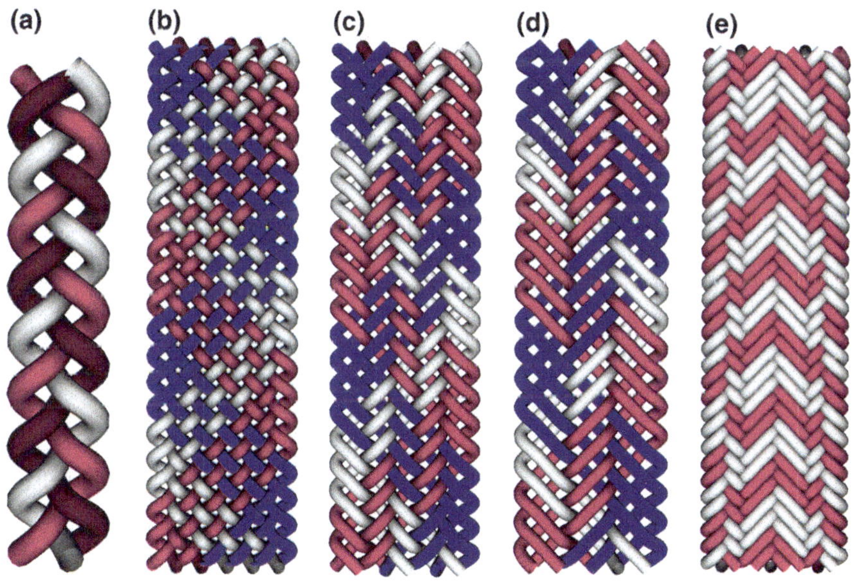

Fig. 2.6 Simulated flat braids **a** with three yarns, **b** with 11 yarns and floating length of one, **c** with 11 yarns and floating length of two, **d** with 10 yarns and floating length of three, **e** fancy braid with floating length 2:2:4:4:2:2

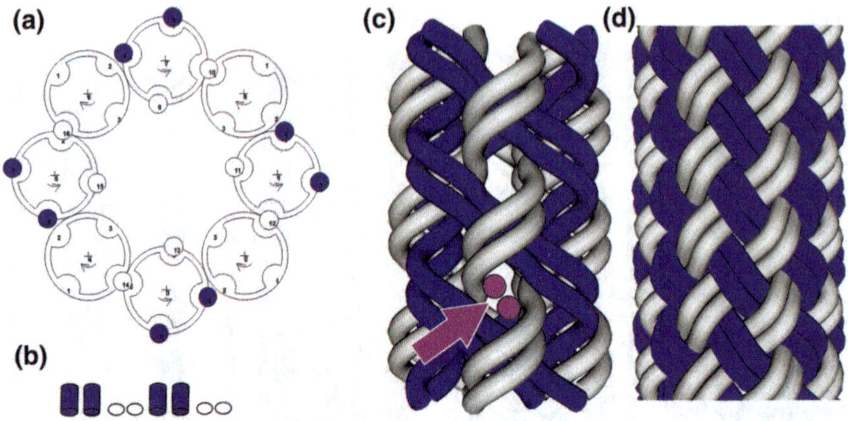

Fig. 2.7 Problem of the topology based algorithm when representing braids with multiple (in this case two) yarns in a group. **a** Braiding machine with occupation 2 Full 2 Empty, **b** the carrier occupation visualizes the two empty carriers, which are reserving empty space, as visible in case (**c**). The braid should look like the configuration (**d**)

also kept for the next two empty spaces, as Fig. 2.7c demonstrates. This empty space is actually too large, because the yarns from the opposite direction are crossing it and therefore require space, corresponding to **one yarn diameter** only to go trough the opening. A corrected sample for this structures is shown in Fig. 2.7d.

2.5.1 Where Is the Problem?

If the carriers on the machine do not carry flexible yarns, but extruders for 3D printing, the structure will be created exactly as presented in Fig. 2.7c. If the carriers carry plastically deformable wire or the braid is created over a core, this figure is as well very close to the reality, but in case of using textile yarns, braided with normal density and without a core, the product looks like the case in Fig. 2.7d. The reason for the difference between the emulative topological model to the reality lies in the movement and rearrangement of the yarn position during the braiding. The appearance of the real braid does not only depend on the topology, but as well on the mechanics of the braiding process, including the acting forces and the mechanical properties of the yarns. These are not considered in the topology based model. In the mechanics of the braiding process—during the braiding, the yarns are under tension and straightened.

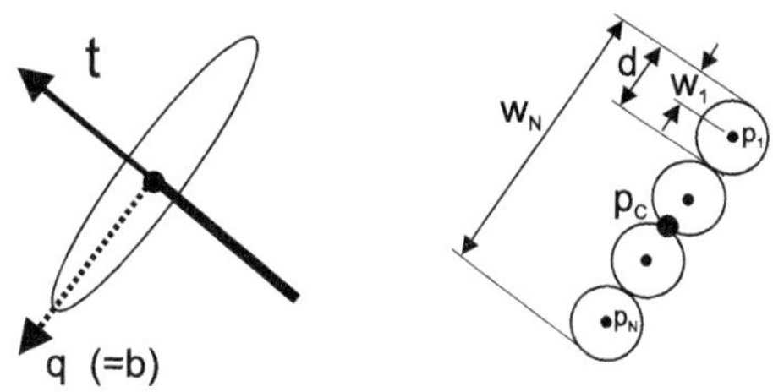

Fig. 2.8 Cross section of the yarn group and its orientation in the space. The orientation (unit) vector **q** is used for the calculation of the coordinates of single yarns in this cross section

2.5.2 *Topology Based Solution of the Mechanical Problem*

In the current case the algorithm can be extended in a way, which can create realistic geometries and the problem can be solved without the application of mechanically based models. The idea is to treat the group as a special case of a multifilament yarn and to apply the algorithm for the creation of multifilament yarn models, explained in the Chap. 9. For this consideration, the cross section of the yarn group is considered elliptical, lenticular or rectangular, with a width w_N and a thickness d. The width of the yarn group depends on the number of the yarns in it.

$$w_n = N \cdot d \tag{2.7}$$

where N are the number of the yarns in the group and d their diameter (Fig. 2.8).

At each point of the yarn axis a tangent, normal and binormal vector can be assigned. If point p_C is assumed to be such a point, it presents the center of the cross section of the group of the yarns as well. Several yarns in one group are oriented in a way, that the moment of inertia of the group section related to the bending axis, is minimal. This orientation corresponds to the minimal bending energy of the system and as well is resulting in no difference in the length of the single yarns, because of the bending. In such a case, the yarn centers have to be located on the bending axis—the line, defined by the vector **q**. It is often parallel to the binormal vector **b**, but in some special cases, like at the edges of the flat braided fabrics, this is not always true.

The coordinates of each filament can be calculated based on the position of the cross section center p_c, orientation of the cross section, based on vector **q**, the total number of filaments N and their diameter d as following:

$$\mathbf{p}_i = \mathbf{p}_1 + \mathbf{q} \cdot d \cdot i \tag{2.8}$$

where the center of the first filament $\mathbf{p}_1$ is calculated as

$$\mathbf{p}_1 = \mathbf{p}_c - \mathbf{q} \cdot d \cdot \frac{N-1}{2} \tag{2.9}$$

Using the Eq. 2.8 all coordinates of the single yarns in the groups, can be calculated as an additional step of the generalized algorithm as following:

- Calculate the **yarn group** axis coordinates according the method described in Sect. 2.4
- Set the Frenet Frame $\mathbf{t} - \mathbf{n} - \mathbf{b}$ for each point
- Define the yarn cross section orientation vector $\mathbf{q}$
- Calculate the point of each yarn based on Eq. 2.8.

The calculation of the tangent vector at each point of the yarn axis is a trivial task. For the normal vector $\mathbf{n}$ simplified values can be used in the current task:

- for tubular braids the nominal cylinder around the braid and not to local curve is enough for the orientation $\mathbf{n} = \mathbf{p_C} - (\mathbf{0}, \mathbf{0}, \mathbf{Z_{p_C}})$
- for flat braids the normal vector shows orthogonal to the nominal plane of the braid surface $\mathbf{n} = (\mathbf{0}, \mathbf{0}, \mathbf{1})$.

This simplification does not come along with a loss of accuracy, but is ensuring additional stability and correctness of the calculations, because the normal vector is always defined. On the contrary, using the classical definition for computation of the normal vector, based on the curve axis, leads to problems, because the yarn axis is not always differentiable. The straight yarn pieces, which appear for instance in the case of yarn floating over more than one yarns, have zero curvature $\kappa = 0$ and on such places the normal vector, calculated using the common equation from the differential geometry,

$$\mathbf{n} = \frac{1}{\kappa} \cdot \frac{\partial \mathbf{t}}{\partial s} \tag{2.10}$$

is not defined. This problem is solved using the predefined normals for the entire structure.

The cross section orientation vector $\mathbf{q}$ can then be defined as vector product $\mathbf{q} = \mathbf{n} = \mathbf{t} \times \mathbf{n}$ in the normal cases. For some special flat braids the situation is explained in the following section.

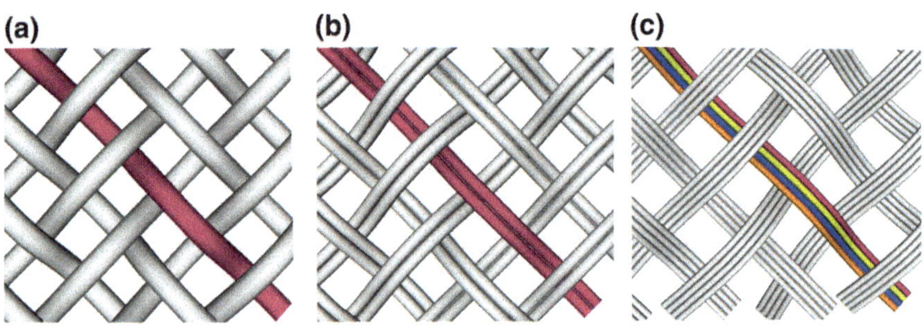

Fig. 2.9 Parts of braid with floating length of two, simulated with the extended (two-step) algorithm for multiple yarns in a group. **a** Basic configuration with single yarn in the group, **b** two and **c** four yarns in a group

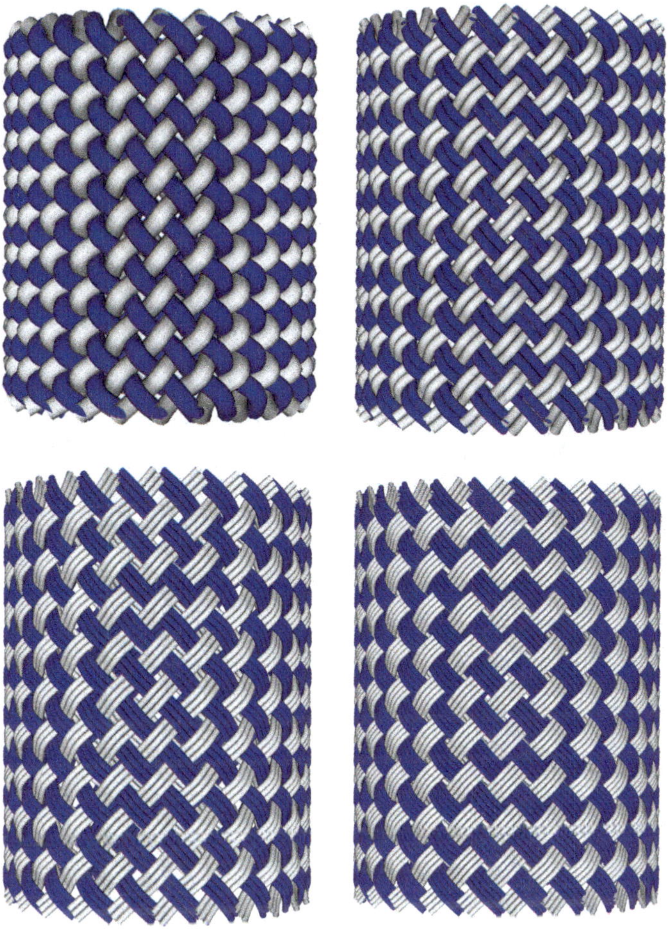

Fig. 2.10 Simulated tubular braids with 48 groups with one, two, three and four yarns in each group

Applying this extended (two step) algorithm, braids with more yarns in a group can also be presented close to the reality. Figures 2.9 and 2.10 demonstrate the application of the algorithm for different structures.

Figure 2.11 presents a modelled tubular braid with a floating length of two and six wires in each group and the photo of such a braid in form of a water hose, in which it is visible, that the simulated geometry is very close to the real one.

Fig. 2.11 Tubular braid with floating length of 2 (regular braid) and 6 wires in a group, used for water hose reinforcement as simulation and photo. Kyosev, Yordan, Generalized geometric modelling of tubular and flat braided structures with arbitrary floating length and multiple filaments, Textile Research Journal, Vol: 86, issue: 12, page(s): 1270–1279, July 1, 2016, Copyright © 2016 SAGE. Reprinted by permission of SAGE Publications https://doi.org/10.1177/0040517515609261 [9]

2.5.3 Flat Braids with Multiple Yarns in a Group

For tubular braids, each group of yarns is following one direction. Once defined, the vector frame $\mathbf{t} - \mathbf{n} - \mathbf{q}$, which is used to define the positions of the single yarns in the group, follows the yarn axis.

For the flat braids this is not the case, because they are build from **one** system of carriers. In this system, each carrier moves half the time in one direction and places a yarn, for instance under angle α. After that, it is reaching one end gear and it is changing the moving direction, placing the yarn under angle $-\alpha$. In the case of multiple yarns being in a group, this process is more complex—and the braids have to be divided into two groups. The first group is the common one—"normal" or "standard" flat braids. In this group the first yarn in the group remains first all the time, as visible in Fig. 2.12a. The second group is seldom used for textiles, but often used for leather and fashion products—it is using a special device named "Wendeklöppel" in German and which translates as "turning carrier". This device carries multiple carriers (instead of one carrier with all yarns) and moves and rotates all them around each gear. This way, the first yarn in the group becomes the last one, as presented in Fig. 2.12b.

For the definition of the moving frame, $\mathbf{t} - \mathbf{n} - \mathbf{q}$ the situation of the "turning carriers" is much simpler—all the time (except the edges!) the orientation (unit) vector $\mathbf{q} = \mathbf{b}$ is equal to the binormal vector at the current point (Fig. 2.13b). Contrary

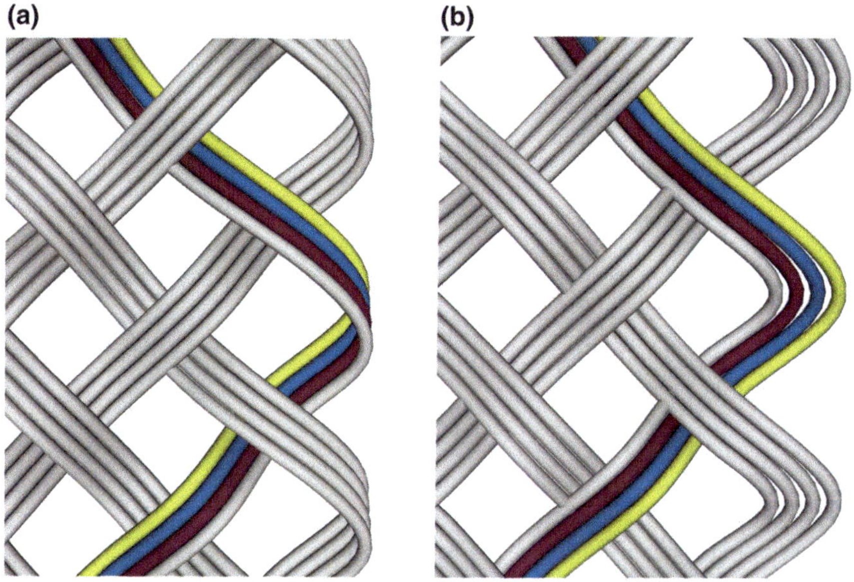

Fig. 2.12 Flat braids with multiple yarns in group can be produced in two ways **a** classical way, where the leading yarn remains leading in both directions, **b** fashion way, using special device (in German "Wendeklöppel"), where the first yarn becomes last after the change of the direction

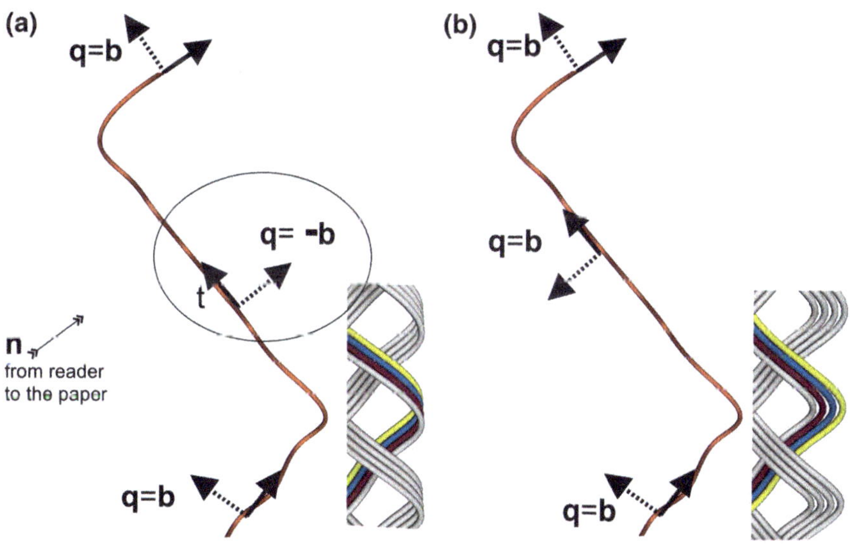

Fig. 2.13 Orientation vector **q** in the "straight" areas between edges **a** for normal braids it is alternating—during the half cycle equal to the binormal vector, the another half cycle opposite; **b** for braids with "turning carriers" equal to the binormal vector

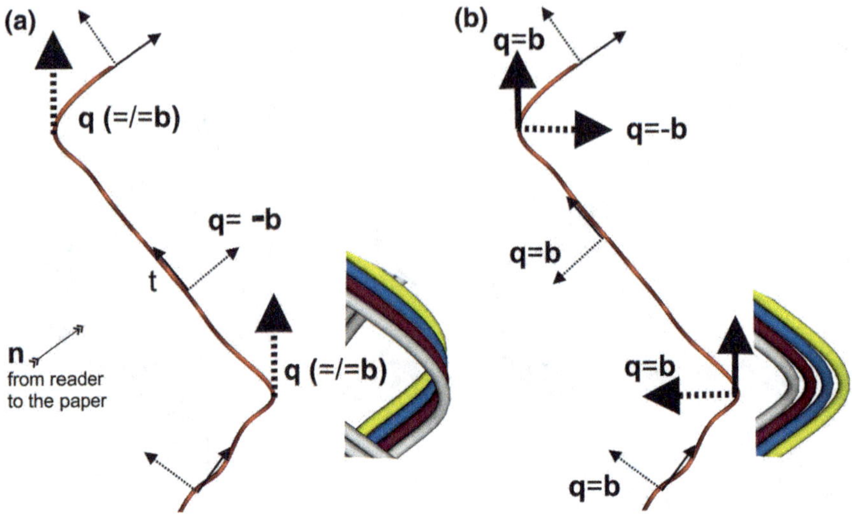

Fig. 2.14 Orientation vector **q** in the areas of the edges. **a** For standard flat braids it becomes parallel to the tangent vector, **b** for flat braids with "turning carrier" it is alternating equal or opposite to the binormal vector

to this, for the most common standard flat braids, the orientation vector is equal to the binormal $\mathbf{q} = \mathbf{b}$ half of the period and the other half, it is opposite to it $\mathbf{q} = -\mathbf{b}$ (Fig. 2.13a).

The placement of the yarn centers in the area of the edges of the flat braid is complex as well. Concerning the "turning carriers", the orientation vector is opposite to the binormal one $\mathbf{q} = -\mathbf{b}$ (Fig. 2.14b) at each left edge, if the tangent vector is defined as a vertical one. The standard braids can actually not use the vector frame at that place, because the orientation vector becomes parallel to the tangent one $\mathbf{q} = \mathbf{t}$ (not to the binormal!), see Fig. 2.14a.

Summarizing the results it can be said that, for the generation of flat braids with multiple yarns in a group ***eight*** different regions are available, four for the "standard" braids and four for the braids with "turning carriers". Each of these regions require separate equations.

2.6 Triaxial Briads

The triaxial braids are braids, in which additional inlay yarns are placed between the yarns of the carriers (named braiding yarns).

The inlay yarns are inserted into the braiding area through a special opening of each horn gear [7].

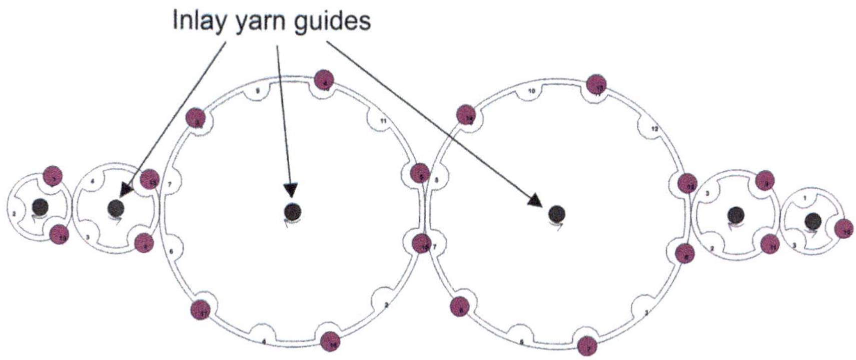

Fig. 2.15 Custom (Fancy) flat braiding machine as basis for generalized model for the placement of inlay yarns in geometrical models. The purple circles are the carriers with the braiding yarns, the black circles are the places, from where the inlay yarns are inserted [23]

Thus, each horn gear can enter only one inlay yarn. In a geometry emulator software, like the TexMind Configurator [8], the modelling of the inlay yarns is simple, because it only requires a placement of inlay yarn on the center of each horn gear Fig. 2.15.

A question more interesting, is the question for the placement of the inlay yarns for generalized braids with arbitrary floating length for geometrical models, which are not based on the process emulation, but on the knowledge about the topology of the structure. For those structures first the main braiding equation is applied [5, 7] in order to determine the floating length of the structure:

$$FL = \frac{N_{SL}}{R_L} \tag{2.11}$$

The repeat length of the carrier arrangement in Fig. 2.15 is two (1 Full + 1 Empty position = 2 carrier positions).

The structure will have six ridges, the first with floating length $FL = 1$, the second $FL = 2$, the third and fourth with $FL = 6$:

$$FL = \left[\frac{12}{2}\right] = 6 \tag{2.12}$$

followed by $FL = 2$ and $FL = 1$. During the time, when one carrier goes around half of the horn gear with twelve slots, its yarn will first cross the yarns of the three carriers, which are currently on the other (upper) half of the side. After that, the three carriers will come to this gear and their yarn will cross as well the yarn of the considered one. Therefore, during the floating of the yarn over the ridge with six yarns, only one inlay can be inserted through the horn gear and it is normally placed in the middle of the crossing area. For the braid with six ridges—six inlay yarns can be placed at a maximum. The coordinates for the placement of the inlay yarn can be

Fig. 2.16 Triaxial flat braid with areas with different floating lengths −1:2:6:6:2:1 [23]

calculated by multiplying the middle distance between two yarn axes with half of the floating length of the ridge.

Figure 2.16 presents simulated triaxial braids which can be produced on the machine, given in Fig. 2.15. It is well visible, that in the large middle ridges only one inlay yarn per ridge is placed, because these areas are build during the motion of the carriers around only one horn gear per ridge. Thus, the possibility for including inlay yarns between each crossing point in some textile modelling software packages is not correct from the braiding point of view.

Figure 2.17b presents a triaxial flat braided structure with a floating length of two (regular braid). Each ridge can only adopt one inlay yarn from the machine. Of course, this inlay yarn can be a ply yarn of two or more single yarns, but this ply

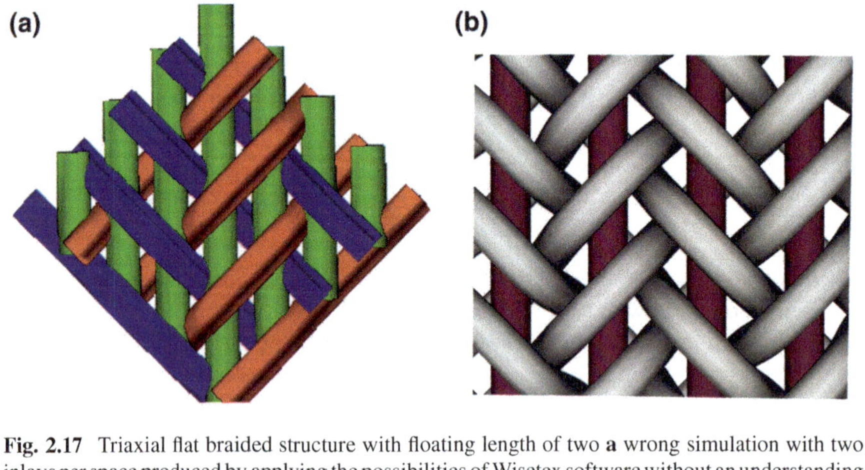

Fig. 2.17 Triaxial flat braided structure with floating length of two **a** wrong simulation with two inlays per space produced by applying the possibilities of Wisetex software without an understanding of braiding **b** correct triaxial braid with floating length of two

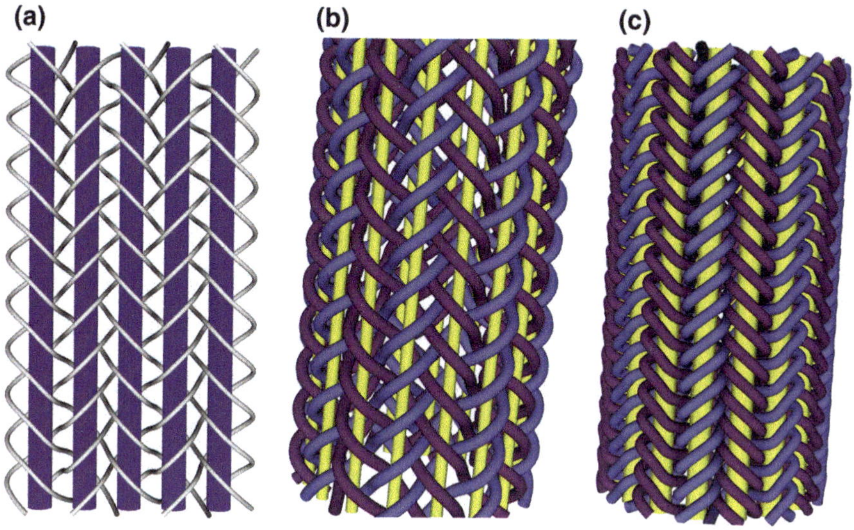

Fig. 2.18 Triaxial tubular braid with floating length of two (**a**, **b**) and three (**c**) [23]

yarn must in any case be placed within the space between the yarns. There is no possibility to distribute two inlay yarns exactly on the line of the crossing points as presented in the Fig. 2.17a, as it can be done for instance using the software Wisetex. In this area, Wisetex gives more freedom to the user—allowing him to place larger yarns and to shift them around the axis. This freedom requires responsibility of the user in understanding the braiding process.

The results of the implementation of the model for tubular braids is presented in Fig. 2.18, in which models of triaxial braids with floating length of two and three, respectively, are presented.

2.7 Configuration for Different than 45° Braiding Angle

The described algorithm is working fine for single yarns, but causes problems when creating dense braids with a braiding angle different than 45° and using multiple yarns in a group, as demonstrated in Fig. 2.19. If the yarn centers of the single yarns in one group, are placed at each cross section point along the binormal vector, as explained in Fig. 2.8, some of them interlace with the yarns from the opposite braiding group, but at a braiding angle of 45°, the algorithm works fine. This means, that the orientation of the group along the binormale vector is an exceptional case, from a more generalized rule. This is not often registered as exception, because the braids with 45° braiding angle (which correspond to 90° angle between the yarn systems) are commonly used. The generalization of the algorithm requires the rotation of the vector $\mathbf{q}$, used for the orientation of the yarns in one group. The new vector $\mathbf{q}^*$ can be

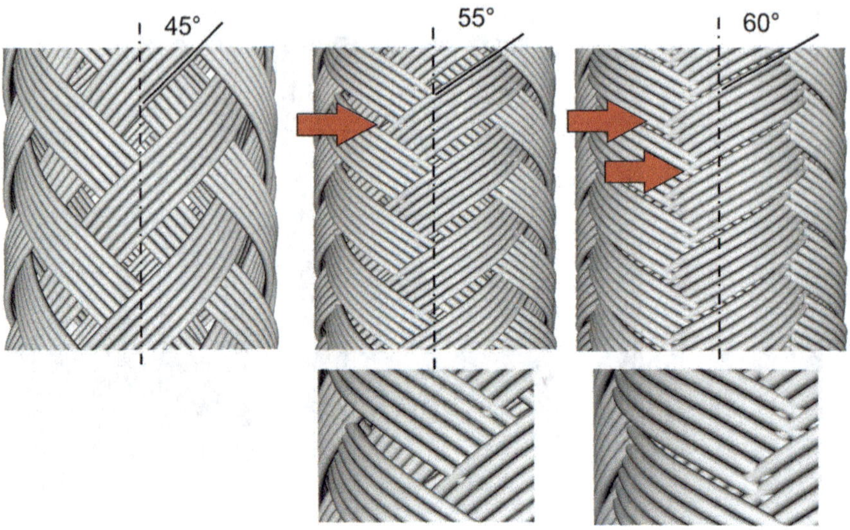

Fig. 2.19 Intersection problem for braiding angles, different than 45°

Fig. 2.20 Placing the orientation vector $\mathbf{q}^*$ for the cross section non orthogonal to the tangential vector, but parallel to the crossing line in order to reduce the interpenetration

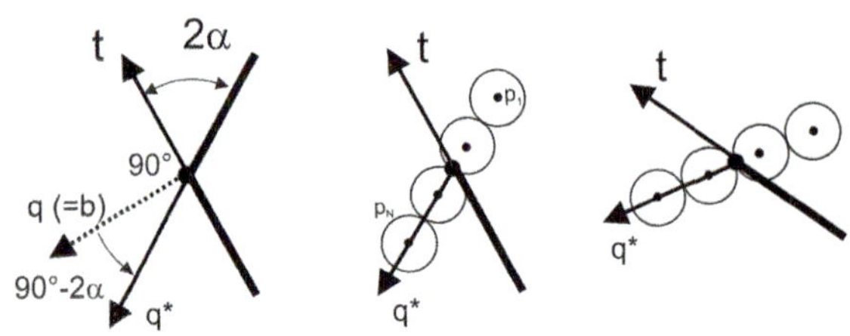

received after the rotation of the (not always binormal) vector $\mathbf{q}$ around the normal vector $\mathbf{n}$ at angle

$$\gamma = 90^\circ - 2\alpha \tag{2.13}$$

as presented in Fig. 2.20.

The rotation of the vector looks trivial, but it has to be performed in the 3D space. One of the simplest ways to perform it, is the usage of transformation matrix [11]

$$R_A(\gamma) = \begin{bmatrix} c + (1-c)\,A_x^2 & (1-c)A_xA_y - sA_z & (1-c)\,A_xA_z + sA_y \\ (1-c)A_xA_y + sA_z & c + (1-c)\,A_y^2 & (1-c)\,A_yA_z - sA_x \\ (1-c)\,A_xA_z - sA_y & (1-c)\,A_yA_z + sA_x & c + (1-c)\,A_z^2 \end{bmatrix} \tag{2.14}$$

where $c = \cos(\gamma)$ and $s = \sin(\gamma)$ and $A = \begin{bmatrix} A_x & A_y & A_z \end{bmatrix}^T$ is the axis of rotation, in the current case this corresponds to the normal vector to the cross section. The normal vector to the yarn cross section for tubular braids is calculated in simple way—from

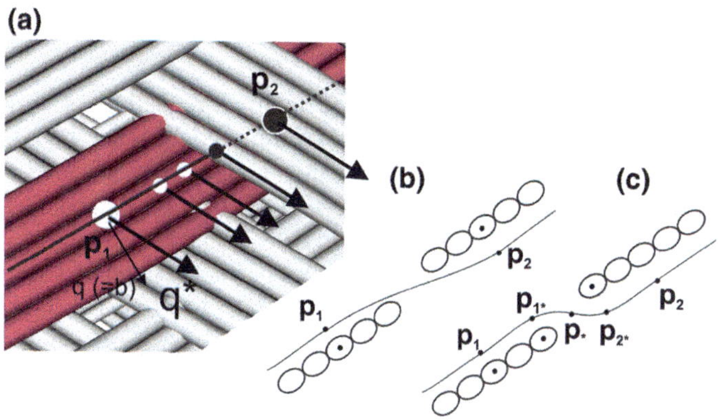

Fig. 2.21 Interpolation and interpenetration problems when using larger group of yarns. Using only the yarn crossing points (**b**) do not provide enough information to keep the geometry clean from interpenetration. More points have to be used for wide yarns (**c**)

the centre of the nominal circle of the current braid cross section $\mathbf{p}_{center} = \left[0\ 0\ p_z\right]^T$ to the current point $\mathbf{p} = \left[p_x\ p_y\ p_z\right]^T$:

$$\mathbf{n} = \mathbf{p} - \mathbf{p}_{center} = \left[p_x - 0\ p_y - 0\ p_z - p_z\right]^T = \left[p_x\ p_y\ 0\right]^T \tag{2.15}$$

The calculated centres of the yarns in a cross section improve the geometry and reduce the intersection for some configurations, but not that significantly, as it is visible in Fig. 2.21a. The problem is, that the centres of the yarns at key points $\mathbf{p}_1$ and $\mathbf{p}_2$ are calculated correctly and distributed correctly, but the configuration of the yarns between these two points is interpolated (Fig. 2.21b). Again, depending on the braiding angle, the yarns on both sides of the groups can intersect with the yarns from the opposite group, because they "meet" them earlier than the central line of the yarn group and the connecting spline at that point is not controlled. The Solution of this problem would require the addition of extra key points on the borders of each yarn group $\mathbf{p}_{1*}$, as for instance $\mathbf{p}_{2*}$ and most suitable in the middle of the space between the yarn groups $\mathbf{p}_*$, so that the curve is constrained and forced to go around the yarn group and not to intersect the single yarns (Fig. 2.21c). Such pure geometrical solution would reduce the intersection between single yarns and will lead to topologically correct structures, so that its implementation definitely makes sense. Still, it will not improve the mechanical correctness of the solution, because the wire materials and the high performance fibers, often used as a group of yarns, are stiffer than the classical textiles and they will not bend around the curve and following it. Resulting from (Fig. 2.21c), these materials will stay as the curve between the cases (b) and (c), causing displacement of the boundary yarns of both yarn groups. The possible solutions of such problems are discussed in a separated chapter.

2.8 Conclusions

The critical review of the works about the geometrical modelling of braided structures, showed a small inaccuracy of some existing models concerning the special case of regular braids. Within the chapter, the mentioned inaccuracy is explained and a solution for its correction is given. A generalized model for the calculation of the yarn paths of tubular and flat braids, using the knowledge of the braiding process, is presented and tested using several different types of interlacements. In addition, the concept is extended for multiple yarns in a group. The implementation of the algorithms in currently only available 3D CAD software in this area "Texmind Braider" and its application for the design of different braids has proven the advantages of the new algorithms.

References

1. Alpyildiz, T.: 3D geometrical modelling of tubular braids. Text. Res. J. **82**(5), 443–453 (2012). https://doi.org/10.1177/0040517511427969
2. Bogdanovich, A.E.: Three-dimensional analysis of anisotropic spatially reinforced structures. Compos. Manuf. **4**(4), 173–186 (1993). https://doi.org/10.1016/0956-7143(93)90002-P
3. Bogdanovich, A.E., Pastore, C.M., Birger A.: Analysis of composite shallow shell structure reinforced with textiles. In: Hamelin, P., Verchery, G. (eds.) Textile Composites in Building Construction, Part 2, pp. 35–44, Paris (1992)
4. Engels, H.: Handbuch der Schmaltextilien. Die Flechttechnologie. Teil I: Maschinen und Verfahren zur Erzeugung konventioneller Geflechte. Mönchengladbach (1994)
5. Goseberg, F., Brabender, K., Moeller, P.: Textiltechnik Maschinengeflechte: Ausbildungsmittel Unterrichtshilfen. Arbeitgeberkreis Gesamttextil, Frankfurt am Mai (1981)
6. Kyosev, Y.: Model generator for tubular braided fabrics. In: Finite Element Modeling of Textiles and Textile Composites, p. 1. Katholieke Universiteit Leuven & St-Petersburg State University of Technology and Design, Saint-Petersburg and Russia (2007)
7. Kyosev, Y.K.: Braiding Technology for Textiles: Principles, Design and Processes: Woodhead Publishing Series in Textiles No. 158, 1 edn. Woodhead Publishing Limited (2014)
8. Kyosev, Y.K.: Machine configurator for braided composite profiles with arbitrary cross section. In: 16th European Conference on Composite Materials ECCM 16, Seville-Spain (2014)
9. Kyosev, Y.K.: Generalized geometric modelling of tubular and flat braided structures with arbitrary floating length and multiple filaments. Text. Res. J. **86**(12), 1270–1279 (2015)
10. Kyosev, Y.K., Brücken, A., Tillmanns, A., Büsgen, A.: Modeling of the geometry and numerical calculation of the elasticity parameters of 3D braided shapes. In: Advani, S.G., Gillespie, J.W. (eds.) Recent Advances in Textile Composites, pp. 255–263. DEStech Publications, Lancaster, Pa (2008)
11. Lengyel, E., Flynt, J.: Math for 3D Game Programming and Computer Graphics, 3rd edn. Course Technology and CENGAGE Learning [distributor], Boston and Independence (2011)
12. Lepperhoff, B.: Die Flechterei, 3rd edn. Leuze Verlag Saulgau/WTTBG, Eugen G (1953)
13. Liao, T., Adanur, S.: A novel approach to three-dimensional modeling of interlaced fabric structures. Text. Res. J. **68**(11), 841–847 (1998)
14. Lomov, S., Gusakov, A., Huysmans, G., Prodromou, A., Verpoest, I.: Textile geometry preprocessor for meso-mechanical models of woven composites. Compos. Sci. Technol. **60**(11), 2083–2095 (2000)

15. Lomov, S., Parnas, R., Bandyopadhyay Ghosh, S., Verpoest, I., Nakai, A.: Experimental and theoretical characterization of the geometry of two-dimensional braided fabrics. Text. Res. J. **72**(8), 706–712 (2002). https://doi.org/10.1177/004051750207200810
16. Lomov, S.V., et al.: Wisetex (2012)
17. Pastore, C.M., Birger, A., Clyburn, E.: Geometrical modelling of textile reinforcements. In: Poe, C.C., Harris, C.E. (eds.) Mechanics of Textile Composites Conference, pp. 597–623. Hampton, Virginia (1995)
18. Proshkov, A.: Mechanisms for Yarn Winding (Topics of Design): In Russian, Original Title: (c), Legpromizdad, Moscow (1986)
19. Rawal, A., Gupta, S., Saraswat, H., Sibal, A.: Geometrical modeling of near-net shape braided preforms. Text. Res. J. **85**(10), 1055–1064 (2015). https://doi.org/10.1177/0040517514559587
20. Rawal, A., Potluri, P., Steele, C.: Geometrical modeling of the yarn paths in three-dimensional braided structures. J. Ind. Text. **35**(2), 115–135 (2005). http://jit.sagepub.com/cgi/content/abstract/35/2/115
21. Rawal, A., Potluri, P., Steele, C.: Prediction of yarn paths in braided structures formed on a square pyramid. J. Ind. Text. **36**(3), 221–226 (2007). http://jit.sagepub.com/cgi/content/abstract/36/3/221
22. Liao, T., Adanur, S.: 3D structural simulation of tubular braided fabrics for net-shape composites. Text. Res. J. **70**(4), 297–303 (2000). http://trj.sagepub.com/cgi/content/abstract/70/4/297
23. Kyosev, Y.: Geometrical modeling of flat and tubular triaxial braided structures. In: The Fiber Society 2016 Spring Conference Textile Innovations—Opportunities and Challenges, pp. 85–86 (2016)

Chapter 3
Evaluation of the Properties of Braided Structures Based on Topological Models

3.1 Introduction

The methods and algorithms of the previous chapter allow the computation of the 3D geometry of yarns in tubular and flat braided structures. This geometry is idealized and does not consider the mechanical properties of the yarns, but it already allows an approximate evaluation of several properties of the braids with useful accuracy. Such parameters are for instance the limits of the free deformation of the braid, named jamming angles, the required length for the production, the weight and the cover factor.

Brunnschweiler developed a geometry based theory about the description of the braiding geometry and the jamming states [2, 3]. Extensions can be found in [4, 5, 19]. Further extensions with applications and verifications for medical braids can be found in [21]. One basic problem, found in all these papers, is that the equation for the cover factor of the braids is derived or re-published in most of the papers, but in none of these, the validity of its range is explained. After an intensive analysis of the equation implemented in the software TexMind Braider [12], the problem was identified and explained in detail in [10]. Experimental verification is reported in [11] and some further results are prepared for publication in [9].

3.2 Relation Between the Braiding Angle and the Elongation of the Braided Products

Figure 3.1 presents a tubular braided structure and its commonly used diameters external—$D_{external}$, nominal or average—$D_{nominal}$, internal or mandrel diameter—$D_{mandrel}$. The yarn diameter is d_f.

In the following equations the nominal diameter $D_{nominal}$ is used. In order to keep the generality of the equations, the nominal diameter will be marked as D without subscript. When the average diameter is replaced with the mandrel or the external

Y. Kyosev, *Topology-Based Modeling of Textile Structures and Their Joint Assemblies*, https://doi.org/10.1007/978-3-030-02541-0_3

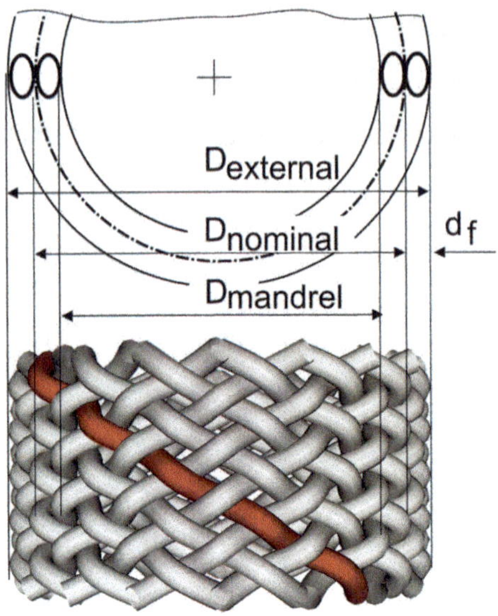

Fig. 3.1 Diameters of a braided structure external—$D_{external}$, nominal or average—$D_{nominal}$, internal or mandrel diameter—$D_{mandrel}$. The yarn diameter is d_f. Reprinted from "Advances in the braiding technologies", Y. Kyosev (Ed.), Y. Kyosev, M. Aurich, Investigations about the braiding angle and the cover factor of the braided fabrics using Image Processing and Symbolic Math Toolbox of Matlab, p. 556, Copyright (2016) with permission from Elsevier [10]

diameter, the values of the investigated parameters can be recalculated, depending on the requested basis of the calculations.

In the following consideration it is assumed, that the braided structure is produced with a diameter D_0 and a braiding angle α_0 (Fig. 3.2) and a length of one repeat of the yarn of h_0. Under axial load the braid can be deformed to another braiding angle and diameter, but the yarn length in one repeat L_f will remain the same.

Based on the relations between the repeat length and the yarn length in both triangles

$$\cos\alpha_0 = \frac{h_0}{L_f} \tag{3.1}$$

and

$$\cos\alpha = \frac{h}{L_f} \tag{3.2}$$

leads to

$$\frac{h}{\cos\alpha} = \frac{h_0}{\cos\alpha_0} \tag{3.3}$$

from there the new repeat length can be found

$$h = h_0\frac{\cos\alpha}{\cos\alpha_0} \tag{3.4}$$

and finally the relation between the braiding angle and the elongation can be achieved

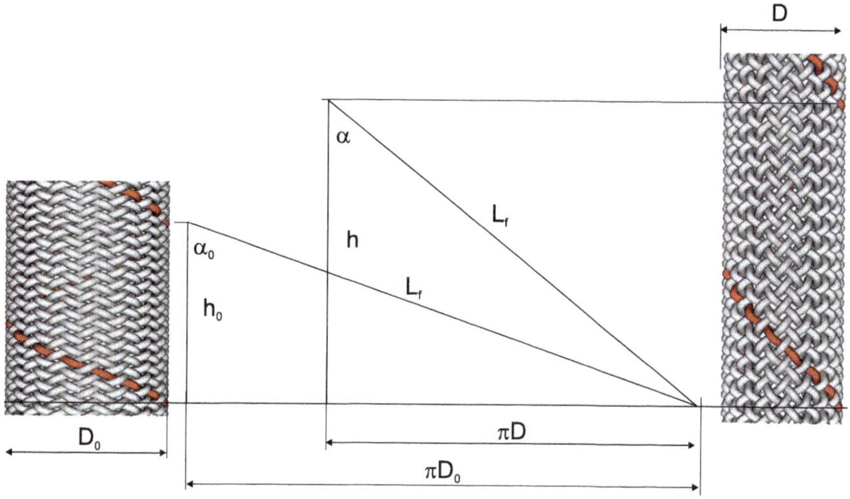

Fig. 3.2 Braided structure in its initial state with diameter D_0 and in elongated state. Reprinted from "Advances in the braiding technologies", Y. Kyosev (Ed.), Y. Kyosev, M. Aurich, Investigations about the braiding angle and the cover factor of the braided fabrics using Image Processing and Symbolic Math Toolbox of Matlab, p. 556, Copyright (2016) with permission from Elsevier [10]

$$\varepsilon = \frac{h - h_0}{h_0} = \frac{h}{h_0} - 1 = \frac{\cos\alpha}{\cos\alpha_0} - 1 \tag{3.5}$$

The Eq. 3.5 can be found in several braiding papers or works related to deformations of springs. It is given here, because it will be required later, for the determination of the elongations of the braid at the jammed states. The braiding angle can be derived from it, using the initial braiding angle α_0 (during the braiding process) and the current elongation of the braided structure ε

$$\alpha = \arccos\left[(1 + \varepsilon) \cdot \cos\alpha_0\right] \tag{3.6}$$

This relation is visualized for $\alpha_0 = 15, 30, 45, 60°$ and $75°$ in Fig. 3.3. This diagram gives a good visualization of the flexibility of the braids. If the braid is produced with a small braiding angle, for instance 15°, there is very small axial elongation possible, but the braid changes its diameter significantly when axially compressed. When a braid is produced with a very large braiding angle (for instance 75°), it has the potential to be elongated, but in the direction of the axial compression it is more close to the maximum limited state. These curves are valid, both, for braided and twisted structures or just for one rod in the form of helix. At this point, the limitations caused by the contact between the yarns based on their interlacement, are not considered. These limitations will not change the curves, they will only change the regions of the curves, which is valid for the braided structure. They are discussed in the section about the jamming states.

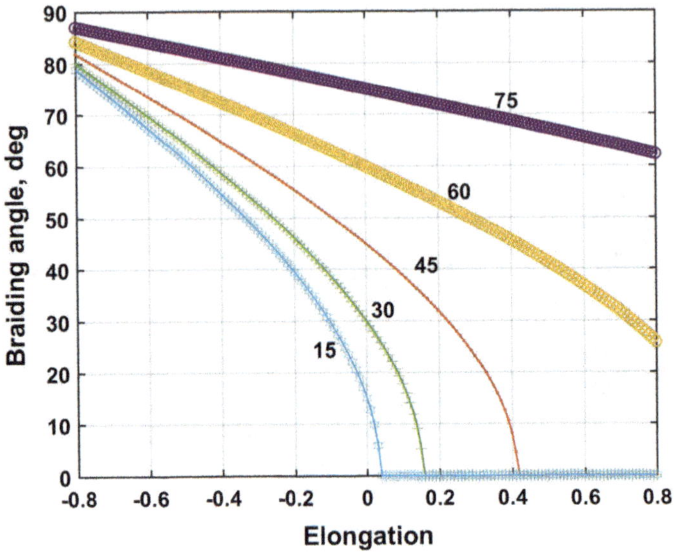

Fig. 3.3 Type of relation curve between the axial elongation or compression of the braid and braiding angle, here without consideration of the jamming limits

3.3 Relation Between the Braiding Angle and the Braiding Diameter During Axial Deformation

Analogically, based on the relation between the circumference of the braid πD_0 and the yarn length L_f

$$\sin \alpha_0 = \frac{\pi D_0}{L_f} \tag{3.7}$$

the relation between the actual diameter D at given braiding angle α can be derived. Taking the relation with the initial values

$$\frac{D_0}{\sin \alpha_0} = \frac{D}{\sin \alpha} \tag{3.8}$$

the current diameter can expressed as:

$$D = D_0 \frac{\sin \alpha}{\sin \alpha_0} \tag{3.9}$$

The relation of Eq. 3.9 is visualized for three different diameters of the braid and the starting angle $\alpha_0 = 45°$ in Fig. 3.4. The relation shows, that with the increase of the braiding angle, the diameter increases as well. Analogically to the diagram in Fig. 3.3 this relation is valid for any one-dimensional continuum with helix form.

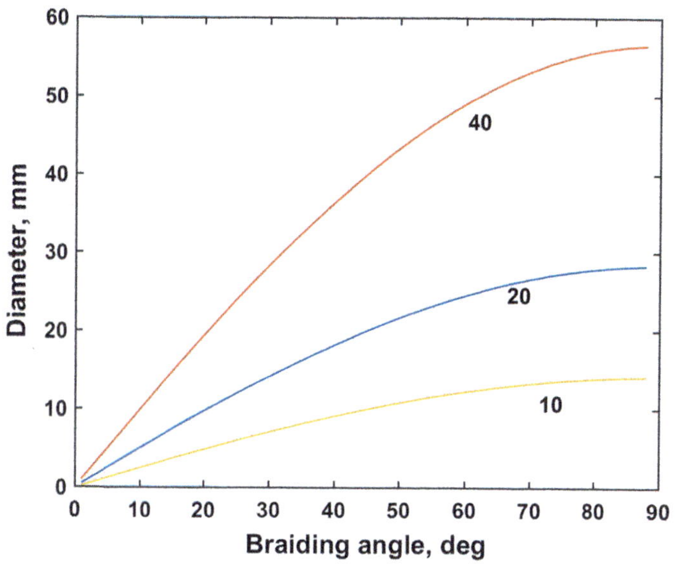

Fig. 3.4 Type of relation curve between the braiding angle and the diameter of the braid, here without consideration of the jamming limits. Initial braiding angle $\alpha_0 = 45°$, initial mandrel (nominal) diameters 10, 20 and 40 mm

Concerning braided structures, only a small range of these curves will be valid, namely the areas, where the yarns can change their orientation before touch their neighbour yarns.

3.4 Unit Cell Geometry

For the current work purpose the unit cell geometry of the braided structure is analyzed at projection level only. Of course, the different pattern will lead to some differences in the freedom of the motion of the yarns—and the braids with larger floating lengths (two or more) will allow the production of denser braids than the ones with normal interlacements. The integration of the consideration of the different patterns significantly increases the complexity of the models, but leads to only small improvement of the accuracy in the boundary areas of the jammed states. Therefore, within this work, the more generalized approach was kept, which allows a quick evaluation of the properties and gives results, which are still of a very acceptable accuracy.

Figure 3.5 presents the unit cell of a braided structure in a projection of a nominal plane for both situations—a braiding angle bigger or smaller than 45°. The situation for $\alpha = 45°$ can be derived from both of these states, but it is less interesting, because it corresponds to an angle of 90° between the yarns, which is equal to the angle in

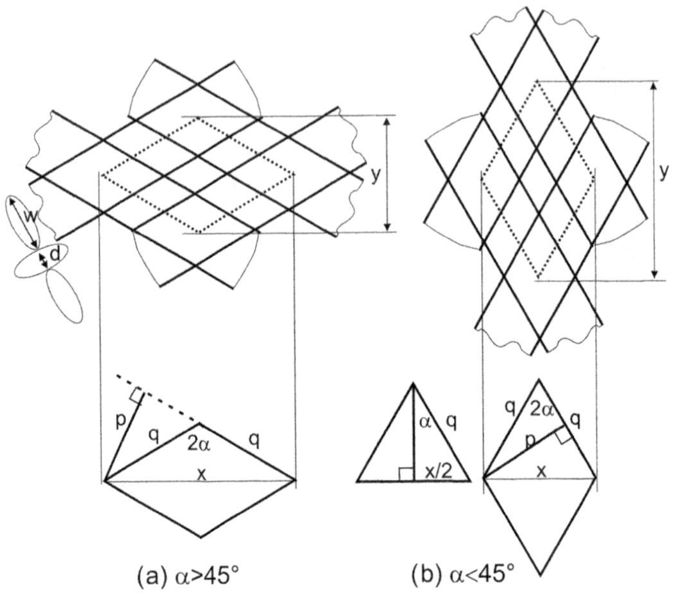

Fig. 3.5 Projection of the unit cell of braids for $\alpha > 45°$ and $\alpha < 45°$ and geometry relations between the width of the cell x, height y, the distance between the crossing points q and the distance between the yarn axes p. Reprinted from "Advances in the braiding technologies", Y. Kyosev (Ed.), Y. Kyosev, M. Aurich, Investigations about the braiding angle and the cover factor of the braided fabrics using Image Processing and Symbolic Math Toolbox of Matlab, p. 558 and p. 559, Copyright (2016) with permission from Elsevier [10]

woven structures. For the unit cells and properties of woven structures, there are several models, papers [14–16] and entire books [1, 18], and therefore, these are not considered here.

A (biaxial) tubular braid with $2N$ yarns (N in $+\alpha$ direction and N in $-\alpha$) has a total number of N yarn projections on the circumference, has N yarns oriented in one direction and respectively N unit cells. The horizontal distance x between two contact points or the width of the diamond will be

$$x = \frac{\pi \cdot D}{N} \tag{3.10}$$

For the side of the unit cell q, which is also the distance between two contact points along one yarn this means

$$\sin\alpha = \frac{x}{2q} \tag{3.11}$$

and the distance p should never become shorter than the yarn width w, but when the yarn thickness d is taken into account:

$$p \geq \frac{w}{2} + d + \frac{w}{2} = w + d \tag{3.12}$$

Between p and q additionally the following is valid (Fig. 3.5)

$$\sin 2\alpha = \frac{p}{q} \tag{3.13}$$

So from (3.13) and (3.12) follows

$$q \cdot \sin 2\alpha \geq w + d \tag{3.14}$$

Replacing q from (3.11) and (3.10)

$$q = \frac{\pi \cdot D}{2 \cdot N \cdot \sin \alpha} \tag{3.15}$$

so finally

$$\frac{\pi \cdot D}{2 \cdot N \cdot \sin \alpha} \cdot \sin 2\alpha \geq w + d \tag{3.16}$$

or, simplified

$$\frac{\pi \cdot D}{N} \cdot \cos \alpha \geq w + d \tag{3.17}$$

Consequently, for a given braided structure, produced with a D, the jamming-free state is defined for braiding angles valid for the following:

$$\cos \alpha \geq \frac{N \cdot (w + d)}{\pi \cdot D} \tag{3.18}$$

The other jamming condition is reached, when the braid is under axial pressure. In this case, according to the frame theory, the frame will have the same geometry but will be rotated at 90° (Fig. 3.6). Therefore, according to this figure the Eq. (3.13) will be changed, but actually

$$\sin(\pi - 2\alpha) = \sin(2\alpha) - \frac{p}{q} \tag{3.19}$$

and as well it is graphically visible, that the cell geometry at the jammed state under tension is rotated at 90° geometry under compression, so the unit cell remains the same, but the angle between the yarns at both states becomes connected through

$$2\alpha_{J(compression)} = \pi - 2\alpha_{J(tension)} \tag{3.20}$$

or

$$\alpha_{J(compression)} = \frac{\pi}{2} - \alpha_{J(tension)} \tag{3.21}$$

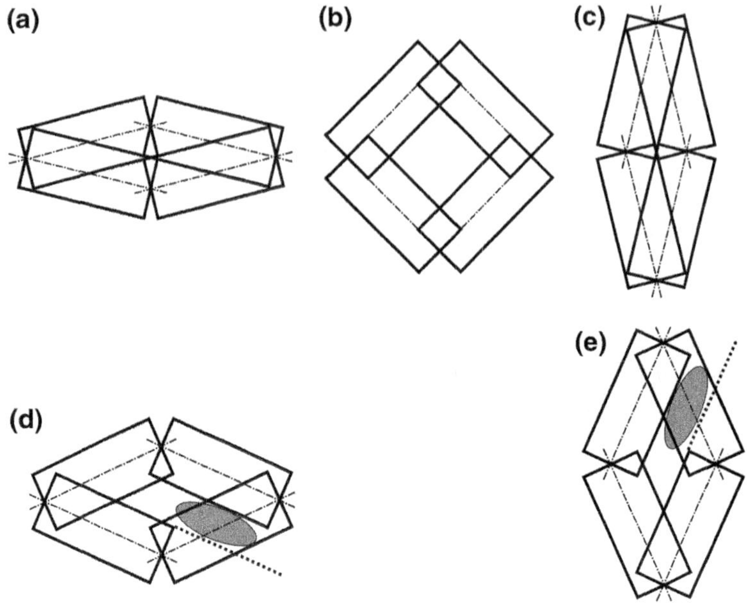

Fig. 3.6 Figure 8. Kinematic model of the yarns in one unit cell in jammed state under axial compression (**a**) and (**d**), (**b**) normal (open) state and (**c**) and (**e**) jammed state under axial tension. Configurations (**d**) and (**e**) consider the yarn thickness. Figure extended on the basis of Fig. 23.8, reprinted from "Advances in the braiding technologies", Y. Kyosev (Ed.), Y. Kyosev, M. Aurich, Investigations about the braiding angle and the cover factor of the braided fabrics using Image Processing and Symbolic Math Toolbox of Matlab, p. 557, Copyright (2016) with permission from Elsevier [10]

This means, that a tubular braid with given diameter D, yarn thickness d and yarn width w and number of yarns in one direction N will only be an **open** structure if the braiding angle is between the values

$$\alpha \leq \arccos \frac{N \cdot (w + d)}{\pi \cdot D} \tag{3.22}$$

and

$$\alpha \geq \frac{\pi}{2} - \arccos \frac{N \cdot (w + d)}{\pi \cdot D} \tag{3.23}$$

Actually the diameter of the braid D in these two situations will not be constant, it will change. If the relation (3.9) remains valid, applying it in (3.22) will lead to

$$\cos \alpha = \frac{N \cdot (w + d)}{\pi \cdot D_0 \cdot \frac{\sin \alpha}{\sin \alpha_0}} \tag{3.24}$$

and applying rules for the sin/cos of the half angle follows for the jamming angle

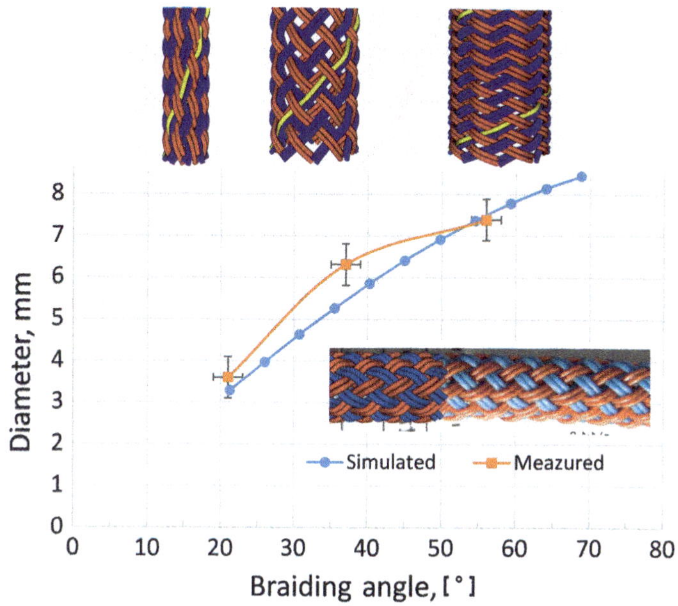

Fig. 3.7 Comparison of simulated and measured jammed states of a tubular braid with 48 single monofilaments in pattern 1:1-2 (floating length of one in a group of two). Reprinted from [11]

$$\alpha = \frac{1}{2} \cdot arcsin \frac{2 \cdot N \cdot (w + d) \cdot sin\alpha_0}{\pi \cdot D_0} \tag{3.25}$$

Popper [17] derived equation (3.25) for tubular braids. Its implementation in the software and combination with the Eq. 3.9 for the relation of the diameter of the braid to the braiding angle allows quick checking of the type of braid and the complete range of its cinematic based deformation. The simulated curve with the jamming limits for a tubular braid is presented in Fig. 3.7. The measured points build a curve with some deviations to the simulated one, but these deviations are caused by the errors of the measurement. The braid is relatively thin (between 3 and 8 mm), but is produced from a stiff mono filament yarn with a thickness of 0.8 mm and due to the accuracy of the measurement, the diagram shows the errors in the worst case of the disturbing factors. For braids with larger numbers of yarns and with larger diameters and softer materials, the accuracy is higher, as demonstrated for instance with the flat braid (Fig. 3.10).

Figure 3.8 summarizes the conditions for achieving an open braided structure or reaching the jamming state in case of axial tension. The diagram presenting the axial compression is symmetrical at the 45° position.

For flat braids, the relation changes slightly—the unit cell depends on the width of the braid and the carriers are not separated into two systems. Half of the yarns are running in one direction and build one group with the angle $+\alpha$ the other half is running in the opposite direction and builds yarn pieces with the angle $-\alpha$. A carrier,

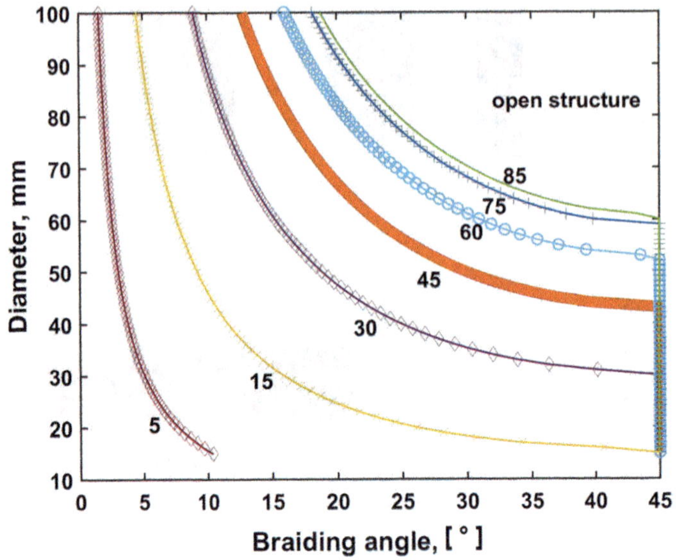

Fig. 3.8 Lower jamming limits (braid under tension), for different combinations of braiding angle and diameter. Each curve represents braid produced at the given initial braiding angle α_0

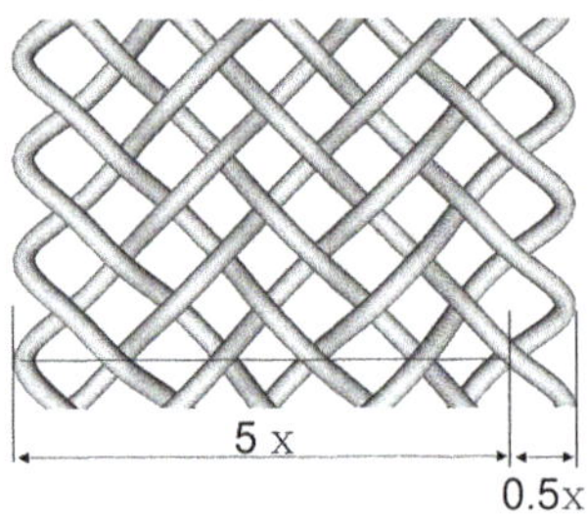

Fig. 3.9 Flat braid with 11 yarns has 5 cells

equipped with the yarn, changes its group from one to another at a defined time. This means, that the number of cells in one braid will be $(N-1)/2$, and for the braid with N = 11 yarns (Fig. 3.9) there will be (11 − 1)/2 = 5 cells. The width of each cell for the tubular braids was $x = \frac{\pi D}{N}$, which in the current case will be replaced by:

$$x = \frac{Width}{\frac{N}{2}} = \frac{2 \cdot Width}{N} \tag{3.26}$$

Replacing it in the Eq. 3.25 for the flat braids, the relation becomes

$$\alpha = \frac{1}{2} \cdot arcsin \frac{2 \cdot (N_{yarns}) \cdot (w + d) \cdot sin\alpha_0}{Width_0} \tag{3.27}$$

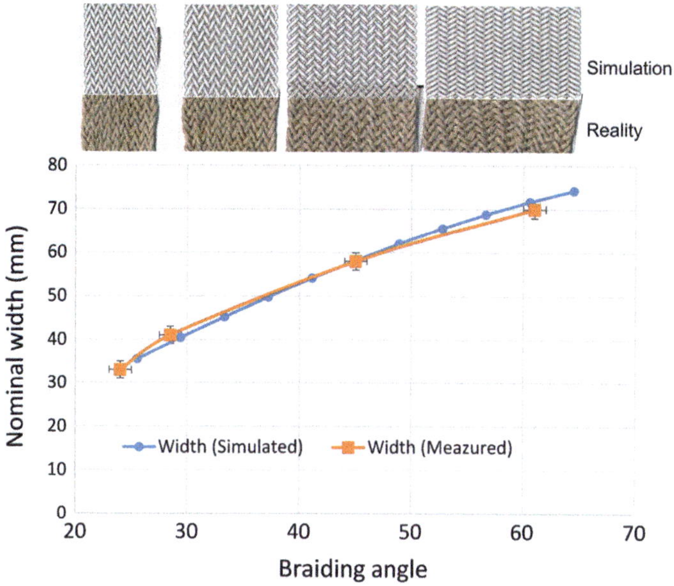

Fig. 3.10 Simulated geometry versus real geometry and "width-braiding angle" relation with the jamming limits for a flat braid with 33 flax yarns with diameter 2 mm, pattern 2:2-1. Braiding angles **a** jamming state under tension at braiding angle 24°, **b** 28°, **c** 45°, **d** practical jamming state under axial compression 61°. Reprinted from [11]

Figure 3.10 shows an optical comparison between the simulated geometry and the real geometry of a braid of 33 flax yarns with a diameter of 2 mm. This yarn was used to check the accuracy of the model for conditions, which satisfy the assumptions of the ideal braid during the creation of the mathematical model yarns with circular cross section. The significantly better correlation between the simulated and the measured value in this case is positively influenced by two factors the size of the braid and the method of the measurement. As the braid has a larger size, the measuring error from the estimation of the width has the same absolute value, but in relation to the measured dimension, this error remains with negligible relative value. Additionally, the width of a flat braid can be scanned and estimated more exact than the diameter of a tubular braid, which is very sensible to touching.

3.5 Yarn Compression and Jamming Limits

Popper [17] derived this equation and also applied the compaction factor K of the yarns, in order to consider the lateral compression of the yarns, when deriving their thickness $d = K \cdot d_{initial}$, in the way, explained in [8]. In order to consider the yarn deformation in this equation, both directions or the cross section change have to be considered, as it is done for woven fabrics in the Wisetex software [14]. The

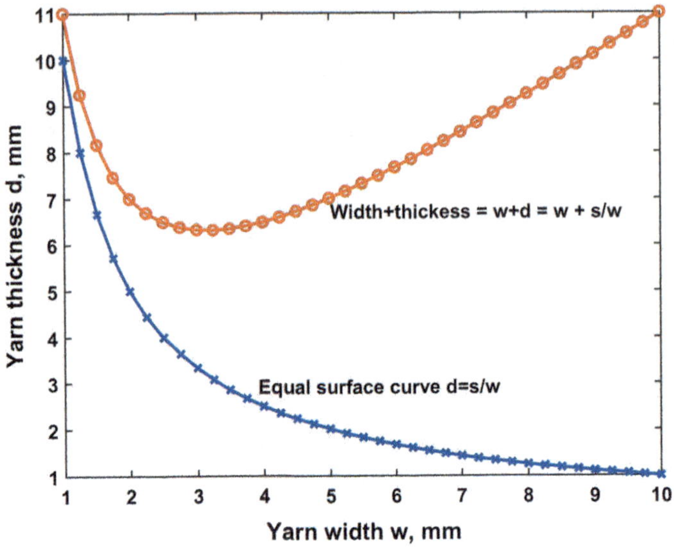

Fig. 3.11 Relation between the yarn thickness and yarn width, in case of constant cross section area and the character of the curve $w + d$, determining the jamming state of the braid. Assumed is surface S = 10 mm^2

measurement of the lateral compression of the yarns and its relation to the change of the yarn width is investigated for twisted ply yarns in [6]. For rovings of high performance fabrics this behavior can be of one type, for twisted yarns used for ropes, it can be of a different one.

If the yarn cross section is considered to be close to rectangular (lateral), then, under the assumption that the fiber volume fraction or the yarn compaction remain identical, the surface of the yarn cross section S should remain constant, if the width or the thickness changes $S = w \cdot d$, so if the yarn thickness d is known, the width will be

$$w = \frac{S}{d} \tag{3.28}$$

and the therm in the Eq. 3.25 $w + d$ will be equal to

$$w + d = w + \frac{s}{w} \tag{3.29}$$

Figure 3.11 depicts both Eqs. 3.29 and 3.28 for a heavy roving with a width of 10 mm and a thickness of 1 mm, or a surface of the cross section of 10 mm^2. If the yarn gets pressed laterally, it can reduce its width and increase its thickness, but the sum of these determines the space required for one crossing unit (thickness of one yarn and the width of the other, until they get get in contact). This sum $w + d$ participates in the equation for the jamming states and, as the examples shows, concerning the current case its values are always lower in the middle range than in the boundaries. The minimum of the function $w + d = w + s/w$ will determine the most compact state of the braid, which is the jamming limit with consideration of the yarn compaction in both directions. The minimum of the function $f(w) = w + d = w + \frac{s}{w}$ is at the value of w for which the first derivative is equal to zero:

$$\frac{df(w)}{dw} = \frac{d(w + \frac{s}{w})}{dw} = 1 - \frac{s}{w^2} = 0 \tag{3.30}$$

This equation has two mathematical solutions

$$w = \pm\sqrt{s} \tag{3.31}$$

but a negative width has no physical meaning, so the minimum is the value

$$w = \sqrt{s} \tag{3.32}$$

The minimal value of the sum $w + d$ is then

$$w + d = \sqrt{s} + \frac{s}{\sqrt{s}} = 2\sqrt{s} \tag{3.33}$$

and if the cross section keeps its packing density constant, the jamming angle for the highest compressed state of a braid made from multi filament yarns, will be

$$\alpha = \frac{1}{2} \cdot arcsin \frac{2 \cdot N \cdot \left(2\sqrt{w \cdot d}\right) \cdot sin\alpha_0}{\pi \cdot D_0} \tag{3.34}$$

where $s = w \cdot d$ is the cross section of the yarns. This angle is about 30% lower (resp. for the opposite load higher) than the jamming state at which the yarns of both systems just touch, but do not start to deform their cross section. Figure 3.12 presents braiding limits (jamming angle) under tension, according to the pure geometrical equation (3.25) and after consideration of the compressibility of the yarn cross section, according to Eq. 3.34 for the same braid, as presented in Fig. 3.8, for 32 yarns total, w = 5 mm, d = 1 mm, and a braiding angle of $\alpha = 55°$. The upper right corner of the figure covers the configurations between mandrel (braid) diameter and braiding angle, which leads to an open braid. The red line marks the first stage, in which the complete surface of the braid is covered, but the yarn cross section does not deform. If the diameter or the braiding angle becomes smaller, then braiding is still possible, but only ***if*** the yarn cross section consists of a large number of filaments and if these can move in a way, that they change the form of the cross section to another one. Values for the diameter from the left upper side of the blue curve can not be produced practically. In these cases, the structure is too dense and building a new braid during the braiding process is faster than taking it off of the machine [13].

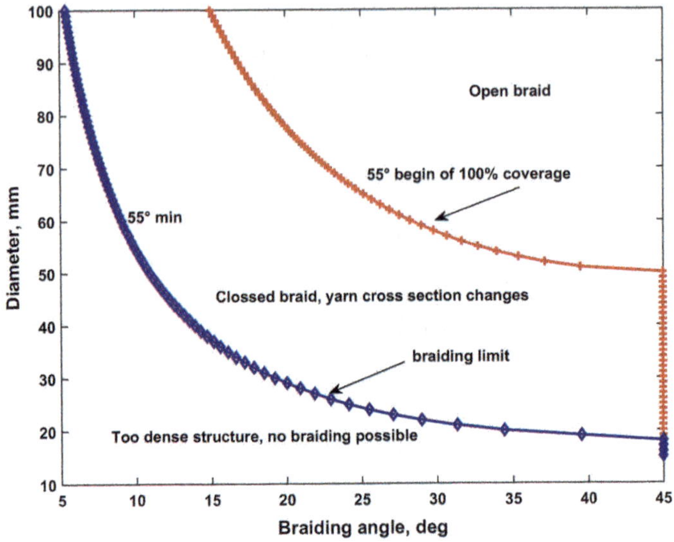

Fig. 3.12 Braiding limits (jammed state) without and with consideration of the yarn cross section compaction. Assumed is a yarn width of 5 mm, a yarn diameter of 1 mm, 32 yarns, and an initial braiding angle $\alpha = 55°$

3.5.1 Notice on the Geometrical Models

There are several works, which describe the crimp of the yarns, depending on the pattern of the structure (diamond, regular braid, etc.). For instance in the paper [20], a condition of Peirce [16] is used as a basis for the calculation of the limit state of the structure and the corresponding jamming angle of woven structures. In the same matter, geometrically, for each type of yarn cross section (circular, elliptical, lenticular) and for each kind of pattern (regular, plain, diamond or named according [13] as floating length from one to N and from one to M yarns in a group), the equations can be derived. The accuracy of these equations will be higher than the approach used here, which is only based on the projection of the yarns. This higher accuracy is normally only theoretical, and higher only in comparison to the generalized **geometrical** model. The compressibility, the bending stiffness of the yarns and their friction significantly influences the jamming state in reality and any effort of creating a more accurate and pure geometric model for specific structures is a loss of time or academic exercise. A creation of such models only makes sense for particular projects dealing with certain structures. The next larger step of the accuracy in a generalized model can only be reached using numerical procedures with implemented mechanical methods, in which the yarn behavior and interactions are considered.

3.6 Cover Factor for Biaxial Braids

In an analogical way the cover factor of the braid can be calculated.

The surface of one unit cell is (according Fig. 3.13)

$$S_{cell} = p \cdot q = q^2 \cdot \sin 2\alpha \tag{3.35}$$

The area not covered by the yarn in the unit cell is (Fig. 3.13)

$$S_{opening} = \left(q - 2 \cdot \frac{w}{2 \cdot \sin 2\alpha}\right)^2 \cdot \sin 2\alpha \tag{3.36}$$

The cover factor (CF) is defined as the ration between the covered unit cell area and the total unit cell area:

$$CF = \frac{(S_{cell} - S_{opening})}{S_{cell}} = 1 - \frac{S_{opening}}{S_{cell}} = 1 - \frac{\left(q - \frac{w}{\sin 2\alpha}\right)^2 \cdot \sin 2\alpha}{q^2 \cdot \sin 2\alpha} = 1 - \left(1 - \frac{w}{q \cdot sin2\alpha}\right)^2 \tag{3.37}$$

Here has to be taken into account, that the Eq. (3.37) make sense only if

$$\frac{w}{q \cdot sin2\alpha} \le 1 \tag{3.38}$$

in other case the Eq. (3.37) gives negative values.

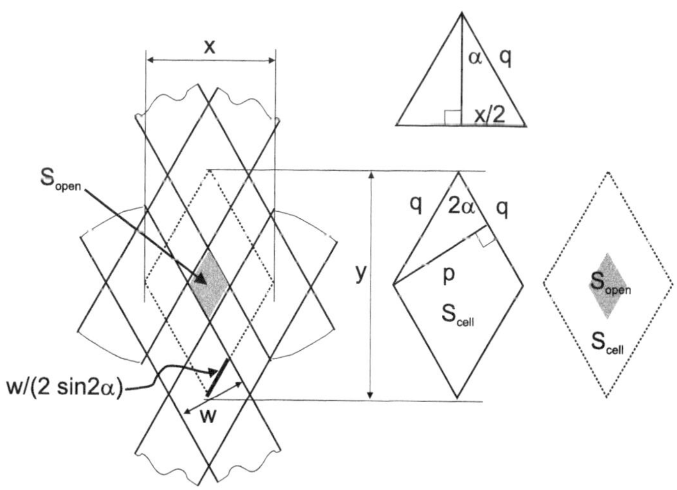

Fig. 3.13 Geometrical relations for the calculation of the cover factor. Figure extended on the basis of Fig. 23.12 of Y. Kyosev (Ed.) "Advances in the braiding technologies", Y. Kyosev, M. Aurich, "Investigations about the braiding angle and the cover factor of the braided fabrics using Image Processing and Symbolic Math Toolbox of Matlab", p. 566, Copyright (2016) with permission from Elsevier

Actually, in this form, the application is not practical, as the value of q can not be measured, if the braid is not produced. Therefore, for the design stage, it make sense to replace the q with the value of (3.15) but this was already done and was determined as the jamming condition in Eq. 3.25:

$$\alpha = \frac{1}{2} \cdot arcsin \frac{2 \cdot N \cdot (w + d) \cdot sin\alpha_0}{\pi \cdot D_0}$$

Replacing q from Eq. (3.15) in the relation for the cover factor

$$CF = 1 - \left(1 - \frac{w}{\frac{\pi \cdot D}{2 \cdot N \cdot \sin\alpha} \cdot sin2\alpha}\right)^2 = 1 - \left(1 - \frac{w}{\frac{\pi \cdot D}{2 \cdot N \cdot \sin\alpha} \cdot 2 \cdot sin\alpha \cdot \cos\alpha}\right)^2$$

leads to

$$CF = 1 - \left(1 - \frac{w \cdot N}{\pi \cdot D \cdot \cos\alpha}\right)^2 \tag{3.39}$$

Note, that N is the number of the unit cells, which is half of the number of the carriers in the tubular braid so

$$CF = 1 - \left(1 - \frac{w \cdot N_C}{2 \cdot \pi \cdot D \cdot \cos\alpha}\right)^2 \tag{3.40}$$

and the correct way of giving this equation looks, contrary to most of the papers, which finish with the equation in the form (3.39), like:

$$CF = \begin{cases} 1 - \left(1 - \frac{w \cdot N_C}{2 \cdot \pi \cdot D \cdot \cos\alpha}\right)^2; & if \ \frac{w}{q \cdot sin2\alpha} \leq 1 \\ 1; & if \ \frac{w}{q \cdot sin2\alpha} > 1 \end{cases} \tag{3.41}$$

If q is unknown, which is common during the design stage, the correct equation is then

$$\begin{cases} \alpha < \alpha_{\min} & CF = 1 \\ \alpha_{\min} < \alpha < \alpha_{\max} & CF = 1 - \left(1 - \frac{w \cdot N_C}{2 \cdot \pi \cdot D \cdot \cos\alpha}\right)^2 \\ \alpha > \alpha_{\max} & CF = 1 \end{cases} \tag{3.42}$$

where

$$\alpha_{\min} = \min\left(\alpha_{Lim}, \frac{\pi}{2} - \alpha_{Lim}\right) \tag{3.43}$$

$$\alpha_{\max} = \max\left(\alpha_{Lim}, \frac{\pi}{2} - \alpha_{Lim}\right) \tag{3.44}$$

and

$$\alpha_{Lim} = \frac{1}{2} \cdot arcsin \frac{2 \cdot N \cdot (w + d) \cdot sin\alpha_0}{\pi \cdot D_0} \tag{3.45}$$

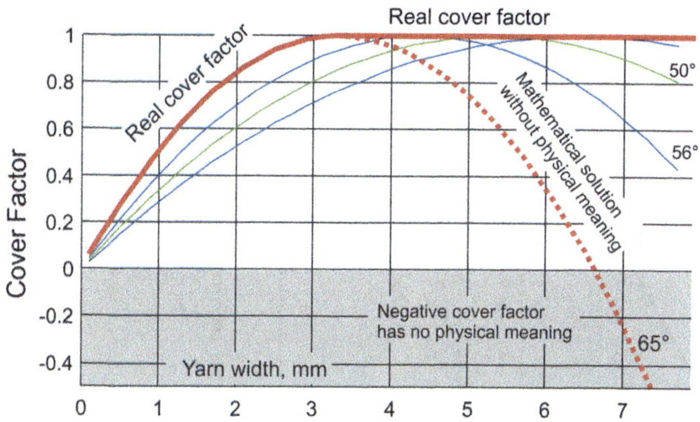

Fig. 3.14 Problem of the cover factor calculation without consideration of the validity of the equation. Only a small part of the calculated values of the commonly known equation, have a physical meaning

Not considering the validity range of the cover factor can lead to serious errors, because the function for the cover factor (3.40) is quadratic and has two branches, which means that there are two solutions as well. Figure 3.14 demonstrates this problem. Calculated was the cover factor of a machine with 48 carriers, for braiding over mandrel with a diameter of 60 mm with a braiding angle between 35° and 65° and for yarn widths between 0.1 and 10 mm. It is well visible, that calculating with smaller yarn widths gives the correct cover factor. By increasing the width of the yarn, the cover factor increases as well. After reaching the jamming state, the equation continues to give *positive* numbers for the cover factors, which are actually *decreasing* and have no more physical meaning. Actually, because these are positive, the risk of considering these as correct values during the design time of the braids, occurs. After increasing the yarn width over some value, the cover factor starts to be negative. This is no more problem from the engineering point of view, because during the usage/process the person, who perform the calculations would recognize, that something is wrong. First using the definition of the state of the braid and then calculating the cover factor, as proposed here, solves these problems. Of course, implementation of the conditions and the equation in a CAD software hides the complexity from the user and allows correct computations, if the software is programmed correctly.

3.7 Cover Factor for Triaxial Braids

For calculation of the cover factor of triaxial braids, the surface of the opening is reduced by the projection of the inlay yarns. The inlay yarns are not always fixed in the middle of the unit cell and their position can influence the cover factor of the

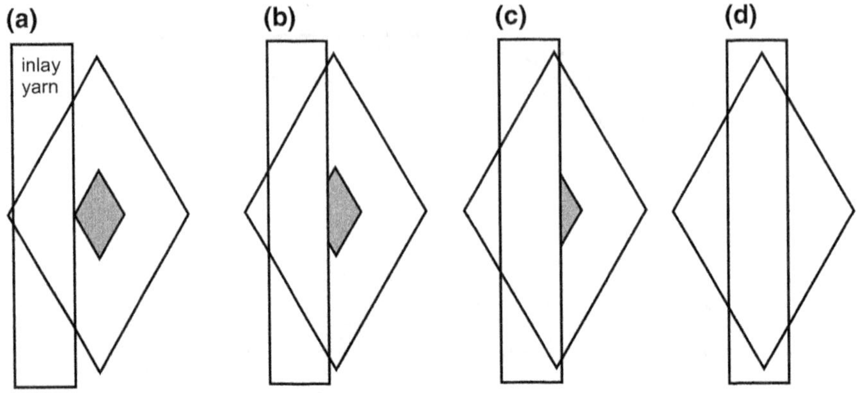

Fig. 3.15 Possible positions of the inlay yarn in a triaxial braid and its influence on the cover factor

braids (Fig. 3.15) significantly. If the inlay yarn is stressed to go to the contact area of the cell, for instance, because of the geometry of the mandrel for the overbraiding or non central position of the braiding point, then it can happen, that it has no influence on the cover factor, because the inlay remains behind the yarns and does not cover the opened area between these (Fig. 3.15a). When after the braiding the inlay yarn can move, it can hide part of the opening (b) and (c). If the inlay yarns have a larger width than the opening between the braiding yarns and they are placed centrally (Fig. 3.15d), they can cover the opening completely and increase the cover factor of the braid to one (or 100%). In the following considerations, this "worst" case will be considered, because in reality the cover factor will always be between the calculated one for the biaxial braid and the worst case.

The surface of one unit cell remains the same (Eq. 3.35), but according to Fig. 3.16 the open area, in this case, will be:

$$S_{opening} = S_{op1} + S_{op2} = 2 \cdot S_{op1} = 2 \cdot \frac{1}{2} \cdot w_{op} \cdot 2 \cdot y_{inlay} = 2 \cdot w_{op} \cdot y_{inlay} \quad (3.46)$$

The surface of one triangle S_{op1} can be calculated based on the width of the open area of one side $w_{op} = 0.5 \cdot (x_{op} - w_{inlay})$ and half of the hypotenuse y_{inlay}, which are connected using through the tangent of the braiding angle

$$y_{inlay} = \frac{0.5 \cdot (x_{op} - w_{inlay})}{\tan \alpha} \quad (3.47)$$

so that:

$$S_{opening} = 2 \cdot \left(0.5 \cdot (x_{op} - w_{inlay})\right) \cdot \frac{0.5 \cdot \left(x_{op} - w_{inlay}\right)}{\tan \alpha} = \frac{\left(x_{op} - w_{inlay}\right)^2}{2 \cdot tan\alpha} \quad (3.48)$$

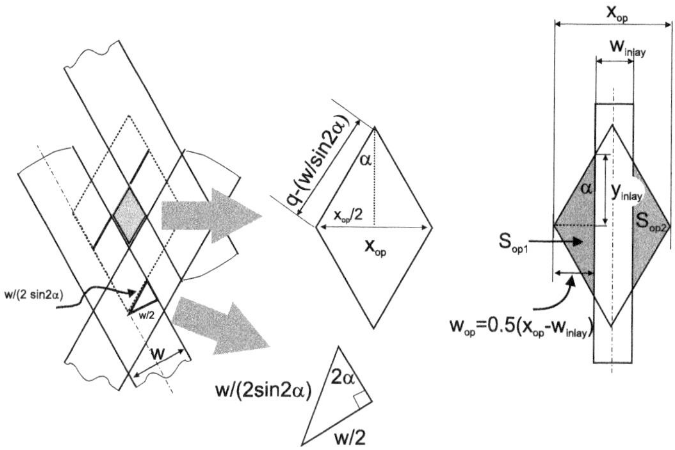

Fig. 3.16 Geometrical relations for the calculation of the cover factor for triaxial braids. Reprinted from Y. Kyosev (Ed.), "Advances in the braiding technologies", Y. Kyosev, M. Aurich, Investigations about the braiding angle and the cover factor of the braided fabrics using Image Processing and Symbolic Math Toolbox of Matlab, p. 565, Copyright (2016) with permission from Elsevier [10]

Remembering that $\sin\alpha = \frac{x}{2q}$ and from Fig. 3.16:

$$\sin\alpha = \frac{x_{op}/2}{q - \frac{w}{\sin 2\alpha}} \tag{3.49}$$

so the width of the opening is

$$x_{op} = 2 \cdot \left(q - \frac{w}{\sin 2\alpha}\right) \cdot sin\alpha \tag{3.50}$$

and the surface of the opening becomes

$$S_{opening} = \frac{\left(2 \cdot \left(q - \frac{w}{\sin 2\alpha}\right) \cdot \sin\alpha - w_{inlay}\right)^2}{2 \cdot tan\alpha} \tag{3.51}$$

The cover factor becomes:

$$CF = 1 - \frac{S_{opening}}{S_{cell}} = 1 - \frac{\frac{\left(2 \cdot \left(q - \frac{w}{\sin 2\alpha}\right) \cdot \sin\alpha - w_{inlay}\right)^2}{2 \cdot tan\alpha}}{q^2 \cdot \sin 2\alpha}$$

$$CF = 1 - \frac{\left(2 \cdot \left(q - \frac{w}{\sin 2\alpha}\right) \cdot \sin\alpha - w_{inlay}\right)^2}{2 \cdot tan\alpha \cdot q^2 \cdot \sin 2\alpha} = 1 - \frac{\left(2 \cdot \left(q - \frac{w}{\sin 2\alpha}\right) \cdot \sin\alpha - w_{inlay}\right)^2}{4 \cdot q^2 \cdot \sin^2\alpha}$$

$$CF = 1 - \left(\frac{2\cdot\left(q - \frac{w}{\sin 2\alpha}\right)\cdot\sin\alpha - w_{inlay}}{2\cdot q\cdot\sin\alpha}\right)^2 = 1 - \left(\frac{2\cdot\left(q - \frac{w}{\sin 2\alpha}\right)}{2\cdot q} - \frac{w_{inlay}}{2\cdot q\cdot\sin\alpha}\right)^2$$

$$CF = 1 - \left(1 - \frac{w}{q\cdot sin2\alpha} - \frac{w_{inlay}}{2\cdot q\cdot\sin\alpha}\right)^2 \tag{3.52}$$

Applying Eq. (3.15) $q = \frac{\pi\cdot D}{2\cdot N\cdot\sin\alpha}$ in (3.52)

$$CF = 1 - \left(1 - \frac{N\cdot w}{\pi\cdot D\cdot\cos\alpha} - \frac{N\cdot w_{inlay}}{\pi\cdot D}\right)^2 \tag{3.53}$$

Which for tubular braid $N = N_c/2$ is the half of the number of the carriers N_c

$$CF = 1 - \left(1 - \frac{N_c\cdot w}{2\cdot\pi\cdot D\cdot\cos\alpha} - \frac{N_c\cdot w_{inlay}}{2\cdot\pi\cdot D}\right)^2 \tag{3.54}$$

Here the conditions are similar—the cover factor can only be between 0 and 1, so the expression in the brackets has to be a number smaller than 1.

To satisfy this condition

$$\left(1 - \frac{N_c\cdot w}{2\cdot\pi\cdot D\cdot\cos\alpha} - \frac{N_c\cdot w_{inlay}}{2\cdot\pi\cdot D}\right)^2 < 1 \tag{3.55}$$

which leads to

$$1 - \frac{N_c\cdot w}{2\cdot\pi\cdot D\cdot\cos\alpha} - \frac{N_c\cdot w_{inlay}}{2\cdot\pi\cdot D} < 1 \tag{3.56}$$

which is always satisfied, because all the variables have positive values and

$$-\left(1 - \frac{N_c\cdot w}{2\cdot\pi\cdot D\cdot\cos\alpha} - \frac{N_c\cdot w_{inlay}}{2\cdot\pi\cdot D}\right) < 1 \tag{3.57}$$

which leads to

$$-1 + \frac{N_c\cdot w}{2\cdot\pi\cdot D\cdot\cos\alpha} + \frac{N_c\cdot w_{inlay}}{2\cdot\pi\cdot D} < 1 \tag{3.58}$$

$$\frac{N_c\cdot w}{2\cdot\pi\cdot D\cdot\cos\alpha} + \frac{N_c\cdot w_{inlay}}{2\cdot\pi\cdot D} < 2 \tag{3.59}$$

So finally for the cover factor it can be set

$$CF = \begin{cases} 1 - \left(1 - \frac{N_c\cdot w}{2\cdot\pi\cdot D\cdot\cos\alpha} - \frac{N_c\cdot w_{inlay}}{2\cdot\pi\cdot D}\right)^2; if & \frac{N_c\cdot w}{2\cdot\pi\cdot D\cdot\cos\alpha} + \frac{N_c\cdot w_{inlay}}{2\cdot\pi\cdot D} < 2 \\ 1 & in\ other\ case \end{cases} \tag{3.60}$$

As mentioned, this equation can give the maximal cover factor, if each inlay yarn stays in the middle of the opening. The minimal cover factor is determined based on the data for biaxial braids.

3.8 Characteristic Points of the Force-Elongation Diagram

The force elongation diagram of braided products depends on a large number of parameters and its prediction is not possible in a simple way, using pure analytical methods. There are several papers, which present theories about the distribution and changes of the fiber tension, depending on the braid load and the braid configuration, but all these papers are based on the assumption, that a relaxed, ready to use braid is investigated. For such a case, they predict the last part of the curve only. Significantly more interesting and important is the prediction of the complete curve. Each braiding company has to produce the braids in some state, then these braids relax, then they go through several treatments as heat setting, stretching and winding, before they are ready to use. Regarding the engineering design it is important to know, how the braid has to look at the beginning—on the machine—in order to have some defined properties in the final state.

The behavior between point 0 and point A in the curve of Fig. 3.17 represents this region and the configuration of the yarns at point A can be predicted, as it is the one of the jamming states of the braids. The reacting tensile force in the braid can only be predicted accurately, if the friction between the yarns, their bending rigidity and their lateral compression, are considered. Theoretically, this can be done well by using FEM modeling, but practically, getting correct and useful results for this region is extremely difficult. The reason for this is, that there are no unified and accurate methods for the measurement of the bending stiffness, lateral compression and friction of fibrous assemblies and, if in some cases these measurements can be done, the configuration of the yarn in the contact places often differs from the configuration during the test. The modeling of the behavior in the region between point A and B relates to the same difficulties, because in this region the movement of the yarn axes is minimal and is only caused by changes of the yarn cross section. If the bending of the yarns can be neglected, the elongation at this limit (point B) can be evaluated by the Eq. 3.34. The last region (B-C) is characterized with predominant tensile loading of the yarn material. The maximal force, which this rope can achieve, will be less than the strength of all yarns, after the consideration of their orientation (Fig. 3.18):

$$F_{braid,\max} \leq N \cdot F_{Yarn,\max} \cdot \cos\alpha \tag{3.61}$$

The prediction of the exact breaking force (Point C) is, again, a more complex issue, because it depends on the variance of the tensile forces of the single yarns of the braid, their load history during the production and finishing processes and the

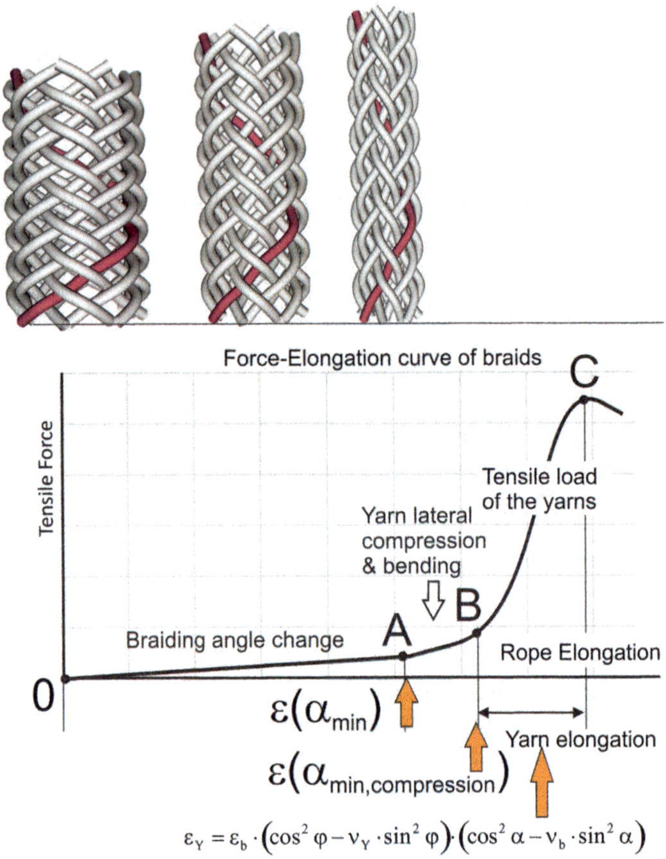

Fig. 3.17 Characteristic points of a force elongation diagram of a biaxial braid [9] with the equation of the relation between the braid and yarn elongation derived in [7]

interaction between all the yarns during the load. For more dense braids, additionally, it has to be taken into account, that the loop strength of the yarns is lower than the normal tensile strength. The elongation at the breaking point and the relation between the elongation of the yarns and the braids, the braiding angle, the crimp angle ϕ and the Poisson number of the yarns ν_Y and the braid ν_b are derived by [7]

$$\varepsilon_Y = \varepsilon_b \cdot \left(\cos^2\varphi - \nu_Y \cdot \sin^2\varphi\right) \cdot \left(\cos^2\alpha - \nu_b \cdot \sin^2\alpha\right) \quad (3.62)$$

Fig. 3.18 Principle relation between the tensile load in the braid and the yarn

3.9 Yarn Length Per Unit Length of Braid

A quick way to evaluate the required yarn length per unit length of a braid is the application of the Eq. 3.1. It does not consider the crimp of the yarns based on the interlacement, but it gives an approximate value for their length. The more accurate geometric models for woven structures have the same disadvantages as the ones used to calculate the jamming angles—these depend on the type of yarn cross section, type of interlacement etc. The created generalized geometry of the braids in Chap. 2 allows an improvement of the calculations. The yarn axis is presented as a curve, which can be described as a function as:

$$f(s) = \langle x(s), y(s), z(s) \rangle \tag{3.63}$$

The length of this curve, respectively the yarn, can be calculated as

$$L = \int_{s=0}^{s_{End}} \sqrt{\left(\frac{\partial x}{\partial s}\right)^2 + \left(\frac{\partial y}{\partial s}\right)^2 + \left(\frac{\partial z}{\partial s}\right)^2}\, ds \tag{3.64}$$

The yarn axis is normally interpolated piecewise, so then the total yarn length is the sum of the single pieces

$$L = \sum_{i=1}^{N} \int_{s=0}^{s_i} \sqrt{\left(\frac{\partial x_i}{\partial s}\right)^2 + \left(\frac{\partial y_i}{\partial s}\right)^2 + \left(\frac{\partial z_i}{\partial s}\right)^2}\, ds \tag{3.65}$$

If a cubic interpolation function is used for each coordinate j and a_j, b_j, c_j and d_j, which are the coefficients of the interpolation function, and s is the curvilinear coordinate for each segment:

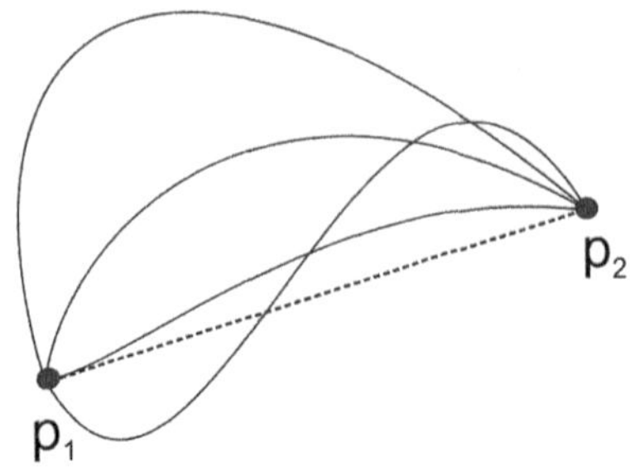

Fig. 3.19 Different paths between two points have different lengths and are always longer than the direct distance between the points, calculated by the Pythagoras resp. trapezoidal rule for numerical integration

$$x_j = a_j s^3 + b_j s^2 + c_j s + d_j \tag{3.66}$$

then the function derivatives can be found analytically and the yarn length of the pieces can be computed analytically:

$$L = \sum_{i=1}^{N} \left[\sqrt{\sum_{j=1}^{3} \left(3\mathrm{a_{ij}} s^2 + 2\mathrm{b_{ij}} s + \mathrm{c_{ij}}\right)^2} - \sqrt{\mathrm{c_{i1}}^2 + \mathrm{c_{i2}}^2 + \mathrm{c_{i3}}^2} \right] \tag{3.67}$$

The alternative approach is the numerical integration for instance using the trapezoidal rule for numerical integration, which in this case is equivalent to the sum of the distances using Pythagoras:

$$L = \sum_{i=1}^{N} \sqrt{(x_i - x_{i-1})^2 + (y_i - y_{i-1})^2 + (z_i - z_{i-1})^2} \tag{3.68}$$

The usage of the Eq. 3.67 or its equivalent for any other interpolation function should be preferred as it gives the exact length of the *interpolating* curve (Fig. 3.19). The Eq. 3.68 on the contrary, gives the shortest distance between the key points only.

Figure 3.20 demonstrates results of a numerical experiment for a braid with 16 yarns(groups) with different floating lengths and different yarns in a group. The simplified equation $1/cos(\alpha)$ gives good prediction of the yarn length but always underestimates this. The length, based on the created 3D geometry, is always higher. The relative error becomes up to 15% for larger braiding angles and is higher for the braids with lower floating lengths (Fig. 3.21), but is less dependent on the number of yarns in a group. All these values are based on the geometry, calculated with pure geometrical models, without contact detection and other mechanical properties. The highest accuracy for the yarn length can be achieved only if the geometry is corrected with such procedures, but they normally require larger computational time and complex algorithms.

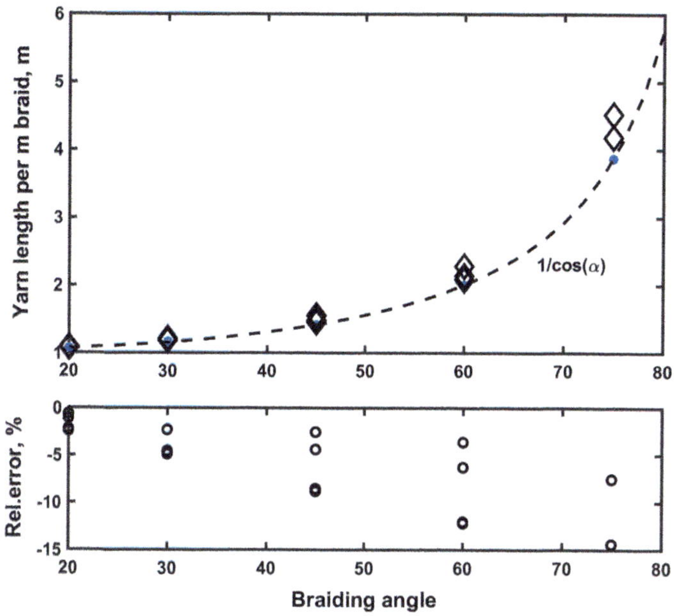

Fig. 3.20 The analytical length of the yarn axis (◇) is always higher than the one of the simplified equation $1/cos(\alpha)$ (presented with dashed line). The relative error increases up to -15% with the increase of the braiding angle

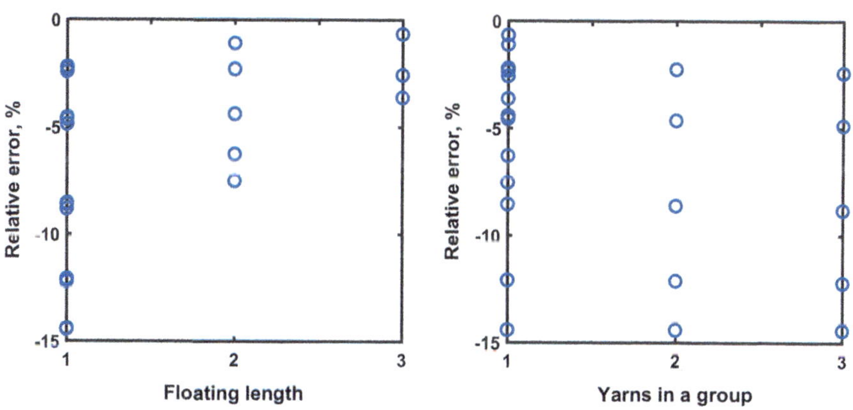

Fig. 3.21 The relative error decreases with the increase of the floating length, as the yarns contain less crimp (left hand side). In addition to that, the relative error does not depend on the number of yarns in the group (right hand side)

3.10 Weight Per Meter

The weight per meter of a braid can easily be calculated when the length of the yarn in the braid is known and the linear density of the yarns is known as well. The accuracy of the weight calculation depends on the accuracy of the length evaluation.

3.11 Conclusions

This chapter demonstrates applications of the topologically generated geometry of the braids for evaluation of the properties of the braided products. The evaluation of the cover factor of biaxial and triaxial braids, jamming states and their elongations as important points of the force-elongation diagram, length per meter and the relation between the braid diameter and braiding angle are discussed. The models and their equations are implemented in software and validated for different configurations. The numerical results demonstrate that the models can provide useful values for all these important for the practical applications parameters of the braids, without their production.

References

1. Behera, B., Hari, B.K.: Woven Textile Structure: Theory and Application. Woodhead Publishing Ltd, Cambridge (2010). http://www.worldcat.org/oclc/610819829
2. Brunnschweiler, D.: Braids and braiding. J. Text. Inst. Proc. **44**(9), P666–P686 (1953). https://doi.org/10.1080/19447015308687874
3. Brunnschweiler, D.: 5—the structure and tensile properties of braids. J. Text. Inst. Trans. **45**(1), T55–T77 (1954). https://doi.org/10.1080/19447025408662631
4. Du, G.W., Ko, F.K.: Unit cell geometry of 3-D braided structures. J. Reinf. Plast. Compos. **12**(7), 752–768 (1993). http://jrp.sagepub.com/cgi/content/abstract/12/7/752
5. Du, G.W., Popper, P.: Analysis of a circular braiding process for complex shapes. J. Text. Inst. **85**(3), 316–337 (1994). https://doi.org/10.1080/00405009408631277
6. Grishanov, S., Lomov, S.E.A.: The simulation of the geometry of two-component yarns. Part I: the mechanics of strange compression: simulating yarn cross-section shape. J. Text. Inst. **88** ((Part 1)(2)) (1997)
7. Hristov, K., Armstrong-Carroll, E., Dunn, M., Pastore, C., Gowayed, Y.: Mechanical behavior of circular hybrid braids under tensile loads. Text. Res. J. **74**(1), 20–26 (2004)
8. Ko, F.K., Pastore, C.M., Head, A.A.: Handbook of Industrial Braiding. Atkins & Peirce (1989)
9. Kyosev, Y.: Simulation driven design of braided products. In: Mahltig, B. (ed.) Textiles: Advances in Research and Applications. Nova Science Publishers, Inc., Hauppauge NY, USA (2017) (in print)
10. Kyosev, Y., Aurich, M.: Investigations about the braiding angle and the cover factor of braided fabrics using image processing and symbolic math toolbox of matlab. In: Kyosev, Y. (ed.) Advances in Braiding Technology. Woodhead Publishing in Textiles, vol. 177. Woodhead Publishing (2016)
11. Kyosev, Y., Cordes, A.: Geometrical modelling of tubular and flat braids within the jamming limits—verification and limitations. In: Kyosev, Y. (ed.) Recent Developments in Braiding and

Narrow Weaving, pp. 23–31. Springer International Publishing, Cham (2016). https://doi.org/10.1007/978-3-319-29932-7_3
12. Kyosev, Y.K.: Texmind braider (2012). www.texmind.com
13. Kyosev, Y.K.: Braiding Technology for Textiles: Principles, Design and Processes. Woodhead Publishing Series in Textiles, No. 158, 1 edn. Woodhead Publishing Limited (2014)
14. Lomov, S., Bernal, E., Koissin, V., Peeters, T.: Integrated textile preprocessor wisetex, version 2.5: computational models, methods and algorithms
15. Lomov, S.V., Verpoest, I.: Model of shear of woven fabric and parametric description of shear resistance of glass woven reinforcements. Compos. Sci. Technol. **66**(66), 919–933 (2006)
16. Peirce, F.T.: 5—the geometry of cloth structure. J. Text. Inst. Trans. **28**(3), T45–T96 (1937). https://doi.org/10.1080/19447023708658809. http://www.tandfonline.com/doi/abs/10.1080/19447023708658809
17. Popper, P.: Braiding. In: Lee, S.M. (ed.) Handbook of Composite Reinforcements, pp. 24–40. Wiley, Hoboken (1992)
18. Postle, R., Carnaby, G., de Jong, S.: The Mechanics of Wool Structures. Ellis Horwood Limited (1988)
19. Potluri, P., Rawal, A., Rivaldi, M., Porat, I.: Geometrical modelling and control of a triaxial braiding machine for producing 3D preforms. Compos. Part A: Appl. Sci. Manuf. **34**(6), 481–492 (2003). https://doi.org/10.1016/S1359-835X(03)00061-7
20. Rawal, A., Kumar, R., Saraswat, H.: Tensile mechanics of braided sutures. Text. Res. J. **82**(16), 1703–1710 (2012). https://doi.org/10.1177/0040517512445340
21. Yuksekkaya, M.E., Adanur, S.: Analysis of polymeric braided tubular structures intended for medical applications. Text. Res. J. **79**(2), 99–109 (2009)

Chapter 4
Process Emulation Based Development of Braided Structures and Machines

4.1 Introduction

The most common uses of braiding are flat and tubular braidings. The tubular braiding is becoming very popular in the form of radial braiding,especially for the covering of mandrels of different forms (overbraiding). The braiding of solid profiles, in the contrary, is less developed and used mainly for gaskets and only single reports and patents about the production of larger solid braids are known in the literature [14]. A literature review of different braiding technologies for three dimensional braiding, including maypole braiding, two step rotary braiding [8, 20], Cartesian braiding, and lace braiding variation [6], can be found in [3]. Very extensive information about the history, the development, 3D braiding machines and principles, can be found in the chapter of A. Bogdanovich in [5]. In the same book, cartesian braiding [4], hexagonal braiding [21] and 3D-Maypole braiding [17] are presented in other chapters as well. For all these techniques, software tools for the emulation of the braiding process are mentioned, but they are only designed for specific machines. Simulation tools for the overbraiding process, based on kinematic relations, are reported by [2, 10, 25], but the method is implemented only for radial braiding machines. The usage of explicitly the FEM Software for the braiding process simulation is reported by several researchers, but initially developed under the supervision of A. Pickett for Radial braiding [19] and for 3D maypole braiding [20]. These systems require a defined law of motion of the carriers, which in case of radial braiding, can be generated for instance using trigonometrical functions or in case of 3D maypole braiding by using the CAB RUN Software of the company Herzog GmbH. The software CAB RUN is based on a rectangular set of horn gears with discrete positions [8, 20], which can be set in a matrix. This method can not be applied for generalized cases of any braiding machine. Therefore, tools, algorithms or methods for the design of new maypole braiding machines for the production of predefined profiles, are not published until now. Patent literature introduces maypole braiding machines for complex profiles, like L-, T- and other profiles [1] or interlock structures [7, 23, 24]. Brookstein proposed in [7] a system of several single basis modules with horn

Y. Kyosev, *Topology-Based Modeling of Textile Structures and Their Joint Assemblies*, https://doi.org/10.1007/978-3-030-02541-0_4

gears, which can be combined in order to receive 3D profiles on maypole braiding machines. All these ideas and drawings are created manually and, in most of the cases, are presented only as tracks around the horn gears, without the carriers. For most of these configurations the carrier arrangement is not tested and the analysis shows, that the usage of large sets of these are not even possible or are working with a highly reduced number of occupied places. The modern CAD systems, like Autodesk Inventor, Creo, Catia etc., already have machine dynamic simulations and are able to simulate the carrier motion, if the complete machine, including track, gears, carriers, carrier foots etc., is modelled. The limitation this simulation shows, is that it only uses a pure physics-based contact detection between the track channel and the carriers foots and that its speed is so slow, that it is practically not applicable in the development. In addition to that, using the CAD system for checking the machine configurations is also inefficient from the development point of view, because the CAD system itself already requires a large part of the machine to be created, in order to test if it works at all.

This chapter presents an effective solution for such problems—giving algorithms, based on two different strategies for computationally efficient carrier motion emulation and testing the machine configurations. The algorithms and their implementation in software allow the designer to create a virtual prototype of a new complex braiding machine in less than a minute and also to test, if the carrier configuration is working.

4.2 Task Definition

As explained before, methods and tools for the design of braided structures with complex cross sections and for the machines used for their production, are missing. The ideal tool for the developers of braids would be one, in which the braid topology can be described and saved in form of some "braid description". Using the braid description, new algorithms can create a machine for it (Fig. 4.1b). This approach requires a suitable way for the topological description of the braids. The methods of the mathematical topology can be used for the description of simplified structures, but for more complex ones—like three dimensional structures—these methods are too abstract and too far from the understanding and the way of thinking of people creating braided structures. The usual method for the description of woven structures uses a number for each interlacing point, which shows how many yarns are below the current warp yarn. For single layer structures, the complete weave patterning is based on an (even) boolean system (up or down). Such a system, containing numbers, can also be applied for braided structures, if some zero level surface can be defined as a basis for the counting. For flat and tubular braids, this will be no problem, but for other, more complex cross sections, this will not be possible, as already explained in [8]. For more complex structures, a list of several zero-surfaces has to be used.

A more simple approach concerning the realization, is the opposite way—the calculation of the form of the braid based on a given, custom defined braiding machine configuration (Fig. 4.1a). In this case, an iterative, "trial and error" process has to be

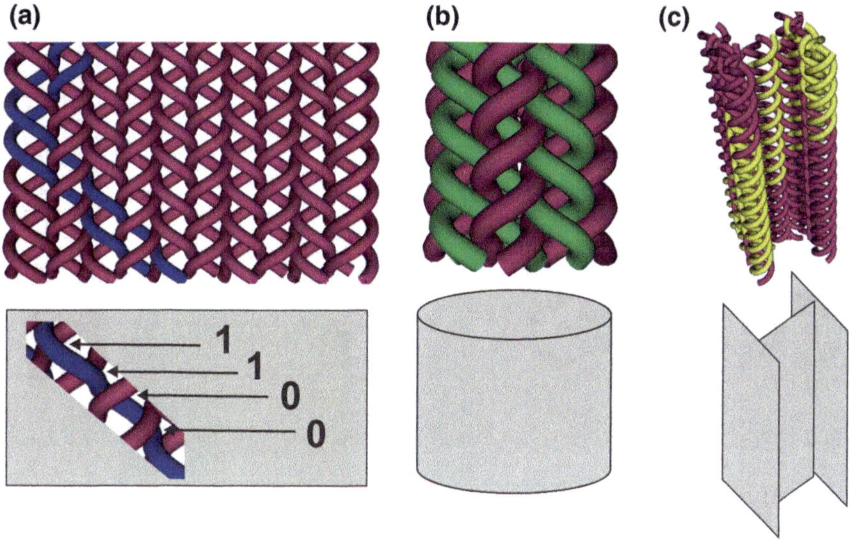

Fig. 4.1 Problem of the normal principle of coding interlacement structures for braids. For flat (**a**) and tubular (**b**) structures the basis surface is one and unique; for more complex structure a set of intersecting basis surface is required

performed several times, adding, replacing, removing or rearranging the horn gears and carriers. Such a tool should virtually emulate the braiding machine and could also be used for the simulation of the complete braiding process at a later stage, if the yarn interaction is integrated into such a model.

In this chapter, the possible steps and methods for the creation of way A—from machine to braided structure—are presented. Some parts of these developments are reported in [11, 13, 15] and the early version of the software implementation is applied in [18]. The parts of the algorithms are implemented in the product TexMind Braiding Configurator [16], which is commercially available (Fig. 4.2).

4.3 Virtual Machine Components

For the creation of a virtual machine, all its important parts have to be modelled with their belonging properties. The modern CAD Systems for the multibody analysis (MSC Adams/Machinery, PTC Creo, Autodesk Inventor etc.) are able to perform a motion simulation of the single parts and to calculate contact surfaces and can be used for performing the task of the modelling of the braiding machine. In fact, these programs are suitable tools for machine engineers, which have to develop the complete construction, calculate the power, the gears, the weights and other parameters. They are oversized and too complex for the usage as a tool for the development and

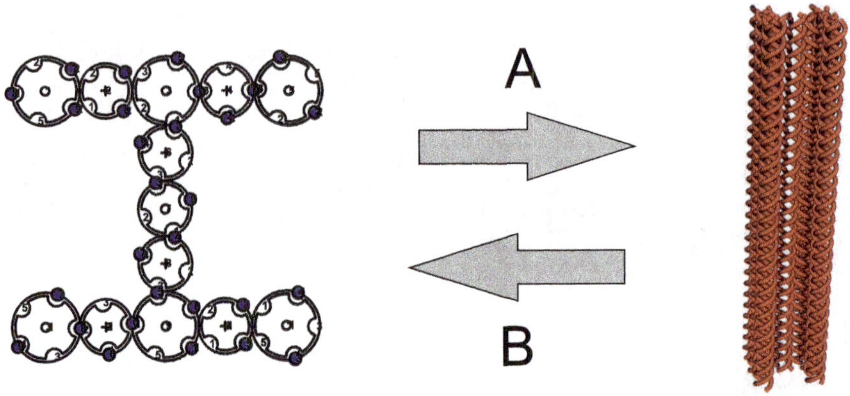

Fig. 4.2 Possible solution ways: A—from machine to structure, B—from structure to machine

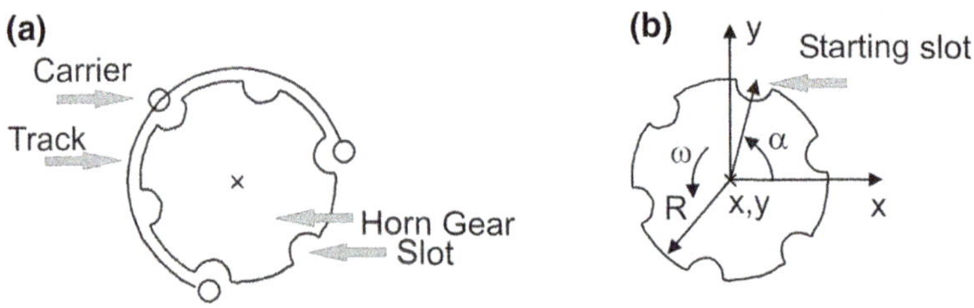

Fig. 4.3 **a** Horn gear, slot, track and carrier, **b** main properties of the object "horn gear"

proof of the **principle configuration** of braiding machines. Therefore, such tools should be implemented within some computational environment, like C/C++, Java, Matlab, Python or others, which can implement all important properties and which are capable of providing a suitable user interface. For the implementation of the Configurator C/C++, programming language was selected and the user interface was realized within wxWidgets library [22]. This guarantees the portability of the library, which means that it can be used on different operation systems.

The virtual objects of a braiding machine are the carriers and the horn gears. For the visualization of the braids, yarns are required as objects. For more advanced technological representations and complex machines, the track (path of the carriers) and its switches can be modelled too.

The virtual carrier (Fig. 4.3) presents the position and the motion of the carriers. It has the following properties as

- Identification number
- coordinates (x, y)
- Pointer or Index of the yarn type.

The carriers can be placed only on the slots of the horn gears (Fig. 4.3), and therefore each virtual horn gear should have the following properties

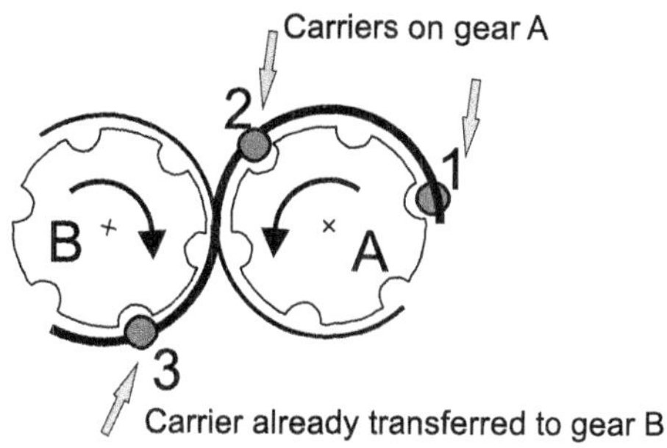

Fig. 4.4 Carriers 1 and 2 are held by the horn gear A, which determines their position. Carrier 3 was initially on horn gear A, but was transferred to horn gear B

- Identification number
- center coordinates (x, y)
- angular velocity ω, which can be positive or negative
- nominal radius R, used for the calculation of the peripheral velocity of the carriers
- current angular position
- local polar coordinate system with reference position (normally the first slot), which is used to determine the positions of the slots
- list of slots, with their angles α in the local polar coordinate systems and with carrier index, which shows the ID of the carrier which is carried or zero, if empty.

When one horn gear is rotating, its current angular position changes. The current coordinates of each carrier C can be calculated based on the current angular position of the horn gear ωt, and the relative angular position of its slot α;

$$\begin{Bmatrix} x_C \\ y_C \end{Bmatrix} = \begin{Bmatrix} x_Z \\ y_Z \end{Bmatrix} + R \cdot \begin{Bmatrix} \cos(\omega t + \alpha) \\ \sin(\omega t + \alpha) \end{Bmatrix} \tag{4.1}$$

After some rotations of the horn gear, some carriers reach an area, where these can be transferred to the next horn gear. The carrier 3 in Fig. 4.4 was initially on horn gear A, but was transferred to B.

The configuration of the horn gears and their transfer places determine the topology of the braided product, because at each transfer place, the carrier and its yarn change their position in the braid—from “inside” to “outside” or opposite—and this way, these determine the interlacement with the other yarns. If the carriers, held by one horn gear, will be transferred to another one, the contact area between these horn gears depends on the track, see Fig. 4.5. The track is a physical channel in which the foots of the carriers are moving [14]. In some cases the track at the contact area between the horn gears is changeable. The contact area from the basis plate is replaced with a small rotatable plate which has both of the channel configurations for transferring and keeping carriers and is named “switch”. All three types of track configurations (transfer, no transfer and switchable) between the horn gears are visualized in Fig. 4.6.

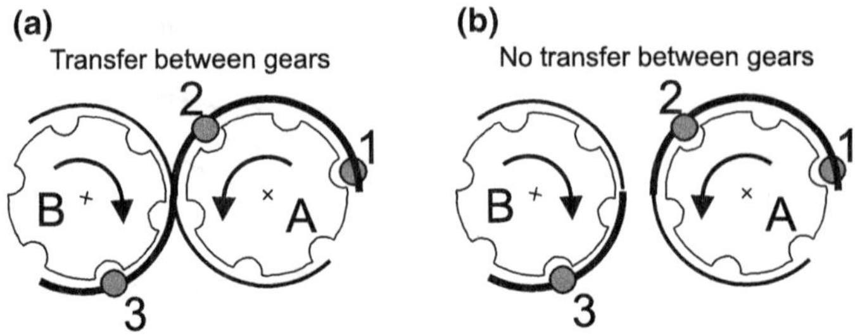

Fig. 4.5 Carriers 1 and 2 are held by the horn gear A, which determines their position. Carrier 3 was initially on horn gear A, but was transferred to horn gear B

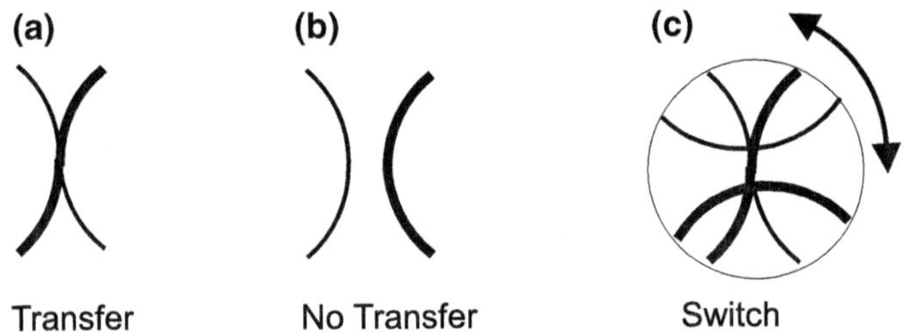

Fig. 4.6 Possible track configurations between two horn gears **a** for transferring carrier to the next horn gear, **b** no transfer—keep the carrier, **c** switch—the function depends on the current angle of the switch, which can be adjusted manually before the machine starts or dynamically using some actuator [15]

In order to model the transfer of the carriers of the braiding machine, these transfer places have to be represented virtually in a suitable way. At this point, there are two different strategies for modelling the complete process, visualized in Fig. 4.7.

Track based modelling The horn gears are used to generate the tracks as a list of curves (Fig. 4.7b). The distance between the carriers is kept constant so the motion of the carriers can be modelled as a motion along the curved paths with constant velocity (Figs. 4.8 and 4.9).

Horn gear based modelling The track is only used to define when the transfer between two gears should take place (Fig. 4.7a). All other steps are performed using rotation of the horn gears and moving the carriers.

Combined (full) model In this case, the combination of the two previously mentioned methods is used.

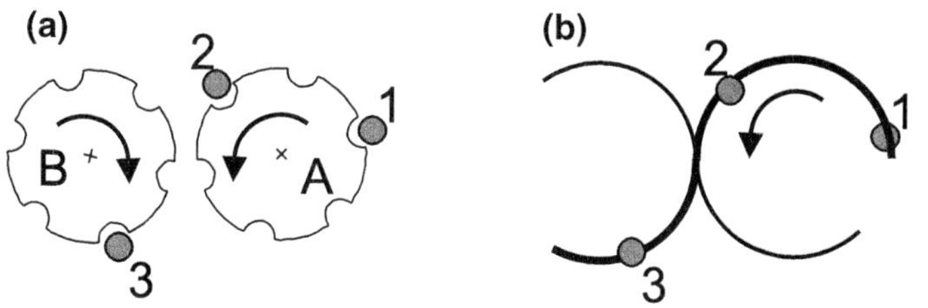

Fig. 4.7 Possible track configurations between two horn gears **a** for transferring carrier to the next horn gear, **b** no transfer—keep the carrier, **c** switch—the function depends on the current angle of the switch, which can be adjusted manually before the machine starts, or dynamically using some actuator [15]

4.4 Track Based Model

4.4.1 Method

The track based model has advantages from the technological and computational point of view. The braiders are handling and configuring the carriers in groups—a group of all carriers in one track moves in the same direction and determines the color appearance for this direction. All carriers in one track receive a yarn, twisted in one direction, the carriers from the other track are twisted in another direction. The track can be separated as a list of connected curves (Fig. 4.11 presents the curve for the machine, visualised in Fig. 4.10), each of it C is defined as an analytical function of the natural parameter s

$$C = C(s) \tag{4.2}$$

The track around the horn gear can usually be presented as a circle. The natural coordinate of one circular arc is connected with the central angle and the radius

$$s = r \cdot (\varphi - \varphi_{Start}) = r \cdot \omega \cdot t - r \cdot \omega \cdot t_{Start} \tag{4.3}$$

and so the circular arc can be defined as

Fig. 4.8 Coordinate system of the carrier motion using track based model

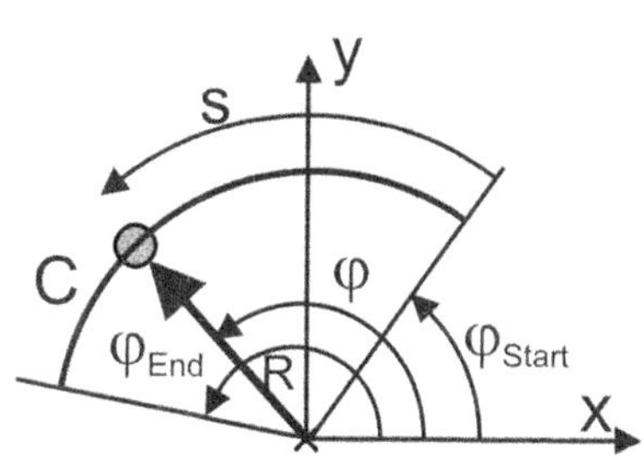

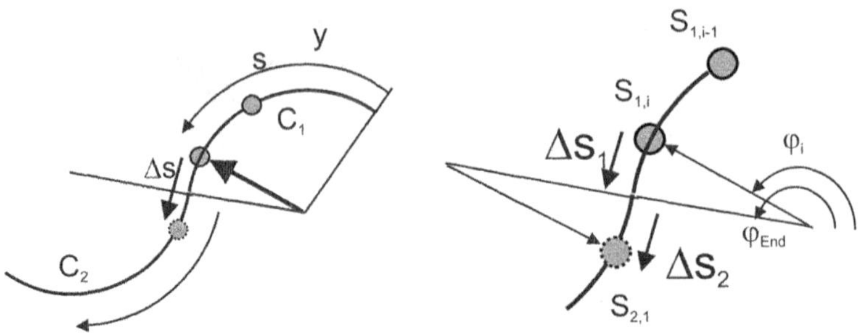

Fig. 4.9 Calculation of the time step and the path at the boundary between two track segments

$$C\left(s\right) = C\left(r \cdot \omega \cdot t\right) = \begin{cases} r \cdot \cos\left(\varphi_0 + \omega t\right) \\ r \cdot \sin\left(\varphi_0 + \omega t\right) \end{cases} \tag{4.4}$$

within the limits (Fig. 4.9)

$$\varphi_{Start} \leq \varphi \leq \varphi_{End} \tag{4.5}$$

Following the Eq. 4.4 the coordinate of the carrier along the curve can be calculated for each time t or angle φ. The carrier changes between the curves, when it is transferred from one horn gear to the other. In case of that during the current time step the path at the current curve is longer than the remaining length of the curve, the carrier continue its motion along the next curve.

The time of the motion around the first horn gear depends on the transfer angle φ_{End}, the angular velocity of the gear ω_1 and the last position of the carrier in angular coordinate φ_i (Fig. 4.9):

$$\Delta t_{1,Last} = \frac{\varphi_{End} - \varphi_i}{\omega_1} \tag{4.6}$$

The path for this time is

$$\Delta s_1 = r_1 \cdot \omega_1 \cdot \Delta t_{1,Last} \tag{4.7}$$

If the simulation time step is Δt, the path on second curve will be

$$\Delta s_2 = r_2 \cdot \omega_2 \cdot \left(\Delta t - \Delta t_{1,Last}\right) \tag{4.8}$$

These equations can be applied for two types of models—with fixed discrete time step $\Delta t = const$ or with variable time step Δt. The variable time step allows computation of the machine state at *every* time moment, if required for some applications. The usage of a discrete time step would only give the state of the machine and the carriers at some predetermined positions.

The motion of the carrier is well determined—it moves with the gear or is transferred to another gear, but a real advantage of keeping the time step variable during the simulation is not recognized. When the time step is fixed to be constant $\Delta t = const$,

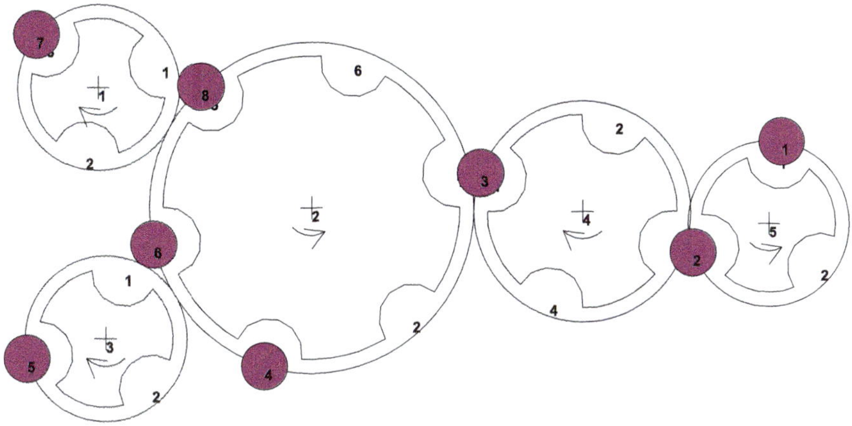

Fig. 4.10 Configuration of braiding machine with 5 horn gears and three sides, used for demonstration of the horn gear based and track based models

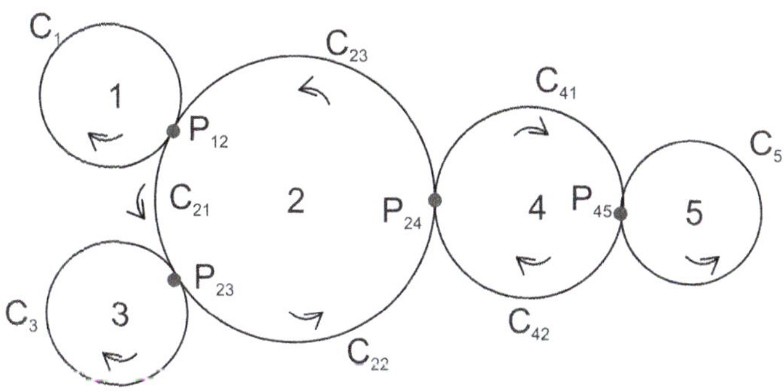

Fig. 4.11 Track based model for emulation of the machine, presented in Fig. 4.10

the model can be simplified through discretisation of the tracks into certain, possible positions, in which the carrier can stay (Fig. 4.12).

In this case the distance between the steps has to consider the circumferential velocity of the carrier at the different gears. In a well designed machine with a high productivity, this velocity is constant and it does not cause problems. The advantage of such a model lies in the simplicity of the time stepping. Once the model is prepared—each carrier has just to be moved to the next one position, placed in the list of all track curves. In order to work, the list of track curves has to be created.

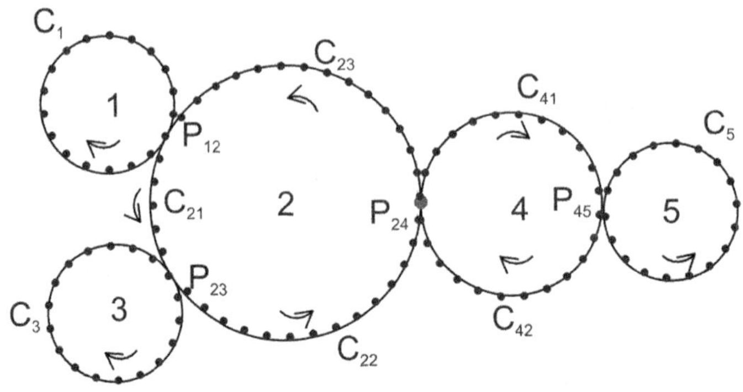

Fig. 4.12 Track based model for fixed time step with discrete allowed positions of the carriers on the tracks

4.4.2 Automated Track Generation

For the simulation of complex machines, the user selects some horn gears and sets their positions. Then the position of the type of the contact interaction between the horn gears, named "contact pairs" (type "interaction"—no transfer of the carriers, or type "transfer" where the carrier moves from one to horn gear to another) has to be defined. After that, the track is created automatically. This can be done in several steps. In the first step, the coordinates of the contacting (transfer) pairs have to be calculated. These are the coordinates of the points P_{12}, P_{23} etc. Fig. 4.13. At these points, all circles around the horn gears have to be split into arcs or general "curves". As demonstrated in Fig. 4.13, only one curve per horn gear will be created around horn gears 1, 3 and 5. Around horn gear number 2, there will be three curves and around horn gear 4—two curves.

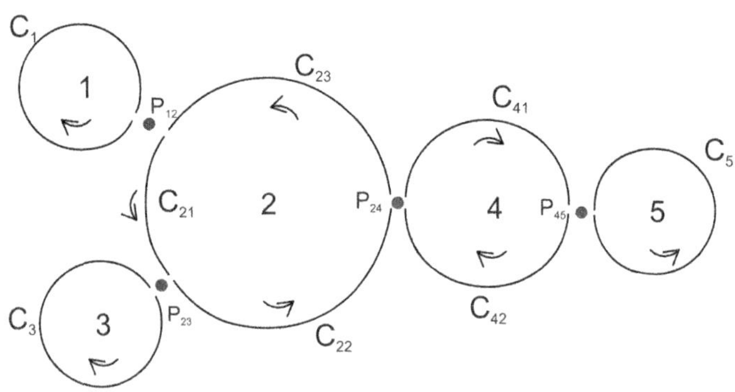

Fig. 4.13 Calculation of the contact point—at which the curves (circle or arc) will be interrupted

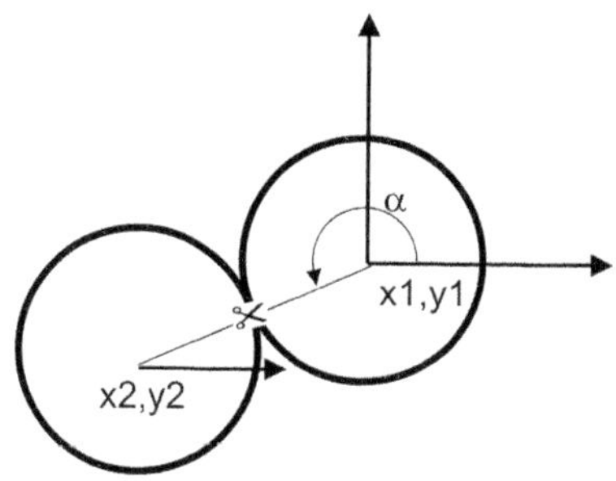

Fig. 4.14 Calculation of the contact point coordinates

The angular coordinate of the contact point between two horn gears can be calculated using simple geometrical relations, as presented in Fig. 4.14.

The angle α can be calculated with the centres of the horn gears (x1, y1) and (x2, y2), using for instance the function $atan2$ which is implemented into most programming languages and which preserves the sign of both y and x components:

$$\alpha = atan2(y2 - y1, x2 - x1) \tag{4.9}$$

The definition of $atan2$ for the current case (where it is not possible, that x and y are equal to 0 simultaneously) is like:

$$\alpha = atan2(x, y) = \begin{cases} \arctan\left({}^{x}/_{y}\right) & x > 0 \\ \arctan\left({}^{x}/_{y}\right) + \pi/2 & x < 0, y \geq 0 \\ \arctan\left({}^{x}/_{y}\right) - \pi/2 & x < 0, y < 0 \\ +\pi/2 & x = 0, y > 0 \\ -\pi/2 & x = 0, y < 0 \end{cases} \tag{4.10}$$

Actually in this form $atan2$ returns value for the angle between $-\pi/2$ and $\pi/2$. For the proper calculation the angle has to be in the range between 0 and 2π, for which the calculated value from (4.9) has changed to

$$\alpha = \begin{cases} a\tan 2(y2 - y1, x2 - x1) & y2 - y1 > 0 \\ 2\pi + a\tan 2(y2 - y1, x2 - x1) & y2 - y1 < 0 \end{cases} \tag{4.11}$$

After splitting the curves into arcs, they have to be inserted into a list of curves, depending on the type of connection points. For the curves in Fig. 4.13 this list will consists of:

- C_1
- C_{21}
- C_3
- C_{22}
- C_{41}

- C_5
- C_{42}
- C_{23}

Depending on the angular coordinate of the carrier, once it is mounted in a slot of a horn gear, the affiliation to the right curves in the list of this carrier, will be determined. From that point on, the carrier will be moved in the slot together with the horn gear until it reaches the end of the current curve. The horn gear, to which the carrier should be transferred later, depends on the information stored in the next curve list. Therefore, the curves do not only take the graphical information about the center, radius and end points into consideration, but they also save the information about their the carriers affiliation to the given horn gear.

4.4.3 Problems with Track Based Models

The implementation of a discrete track based model is not that complex, but it has its disadvantages, especially if it has to be used for machines with some or only motor driven horn gears. Such a machine configuration is the well known 3D braiding machine of the company Herzog [17], the recently developed and smaller variational braider (VF) or the core-sheet-connection machine by the same company. With these machines some gears can move slow or faster or even stop their motion for a certain period of time. In order to simulate such processes, the tracks have to be re-meshed with discrete points every time and some of the gears have to change their speed. The carriers, placed in the track, have to be re-positioned at some of the new calculated positions. This can be implemented, but it is a complex programming procedure, which limits the flexibility of the models.

Another issue is the implementation of machines with dynamic switches (Fig. 4.15. The variable switches can change the type of interaction between two horn gears from the state of "transferring" carriers to the state of "keeping" the carriers. In such a case, both tracks of each switch (at "transfer" and "keep" state) have to be meshed with positioning points. Once the switch changes its state, the list of the curves has

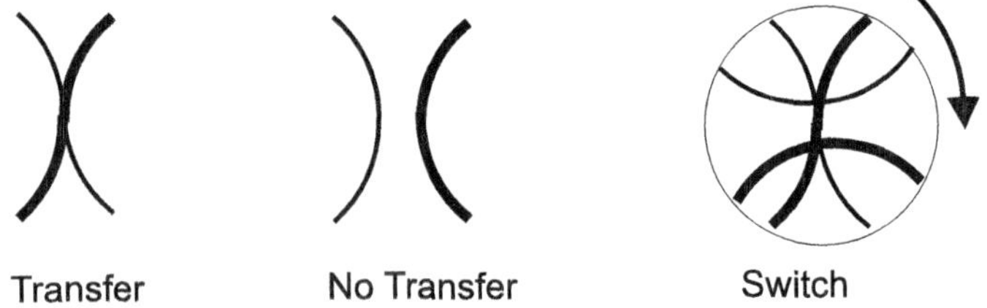

Fig. 4.15 Possible state of "switches" between horn gears, **a** transfer carriers, **b** no transfer (or "keep"), **c** possible realization of the track curves for rotatable switch

to be re-initialized. From the programming implementation point of view this is not a problem, because the list of the curves can be defined dynamically using pointers, so that each end of the curve keeps a pointer (or index) until the beginning of the next curve.

Accordingly, the only problem for the discrete track based models remains the implementation of gears with variable speed. Keeping the track as a curve and not discrediting it onto positions will eliminate this problem, but will require more calculations for the identification of the transfer state. The implementation of this algorithm in the initial version of the software TexMind Braiding Configurator [15, 16] and demonstrates a problem, which is essential for the simulation. The next angular position of a carrier is calculated with a defined accuracy and the computational errors, occurring during the floating point summation, are collected. As investigated by [9, 26] the summation of floating numbers using computers is connected with summation errors. Even though it can be minimized using the approach, proposed by Kahan [9], this error still exists. This leads to accumulation of an error during the calculation of the next position of a carrier and it is already transferred even though it needs to stay on the gear during this step, or vice versa.

4.5 Horn Gear Based Model

The horn gear based model does not need the real track of the machine. The only information needed of the track, are the places of the switches, which show until which position a carrier should stay on the current gear and at which position it should be transferred to the next one (Fig. 4.16). The horn gear radius depends on the number of slots on the horn gear, if the carrier velocity remains constant during the production process. The carrier velocity is equal to the peripheral velocity of the horn gear $V = \omega \cdot R$, which is automatically guaranteed for the tubular braiding machines or generally for the machines, with only one size (same number of slots) of horn gears. If for instance, for flat braiding machines or for some more complex machines horn gears with less or more slots have to be used, in order to turn the carriers in the opposite part of the track, then such horn gears should have a different angular velocity ω and radius R. For example, when a four slot horn gear and a five slot horn gear or a N slots horn gear should satisfy:

$$V = \omega_4 \cdot R_4 = \omega_5 \cdot R_5 = \omega_N \cdot R_N \tag{4.12}$$

If the Eq. 4.12 is integrated during the time of the rotation of one horn gear at the angle between two slots, then the length of the arc between two slots should always be the same

$$L_1 = V \cdot \Delta t = \varphi_4 \cdot R_4 = \varphi_5 \cdot R_5 = \varphi_N \cdot R_N = \frac{2 \cdot \pi}{N} \cdot R_N \tag{4.13}$$

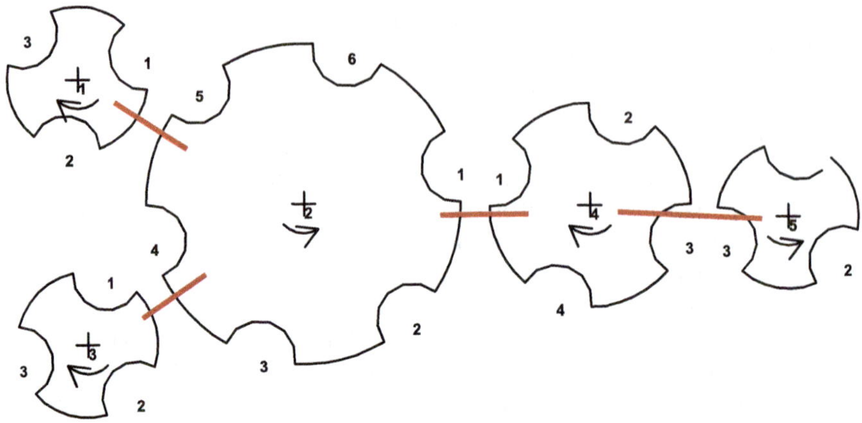

Fig. 4.16 Horn gear based model for emulation of the machine, presented in Fig. 4.8

where the radius of horn gear with N slots have to always be

$$R_N = \frac{L_1 \cdot N}{2 \cdot \pi} = \frac{2 \cdot \pi \cdot R_4}{4} \cdot \frac{N}{2 \cdot \pi} = R_4 \cdot \frac{N}{4} = \frac{L \cdot N}{8} \qquad (4.14)$$

where L is the common used "stitch" of the braiding machine. In the braiding, the "stitch" is the distance between the centers of two horn gears. Actually,today the stitch gives information about the carrier size too. If a horn gear with four slots is chosen, then a carrier for a 80 mm stitch means, that this carrier is designed to suit a machine with horn gears with four slots per horn gear and a distance between the horn gears of L = 80 mm. Once the number N of its slots is known, the radius of each horn gear can be calculated automatically, using the Eq. 4.14.

If a horn gear has to have a different radius or a different speed for a certain time—this is no problem for the horn gear based models, as they carry each of their carriers in the slots, until the transfer position is reached. The realization of switches is also more simple with this approach. If the transfer area between two horn gears has to be set to "keep", this transfer position can be deactivated very easily and the carrier travels further on this gear.

4.6 Automated Carrier Arrangement

Once the settings of the horn gears are created and the tracks of the machine are defined, the carriers can be positioned. One way is, to do this manually by a user, using graphical interface or some predefined file format. More interesting for larger machines can be the possibility of using algorithms for an automated carrier placement. Such an algorithm has to go through the slots, belonging to one track and fill

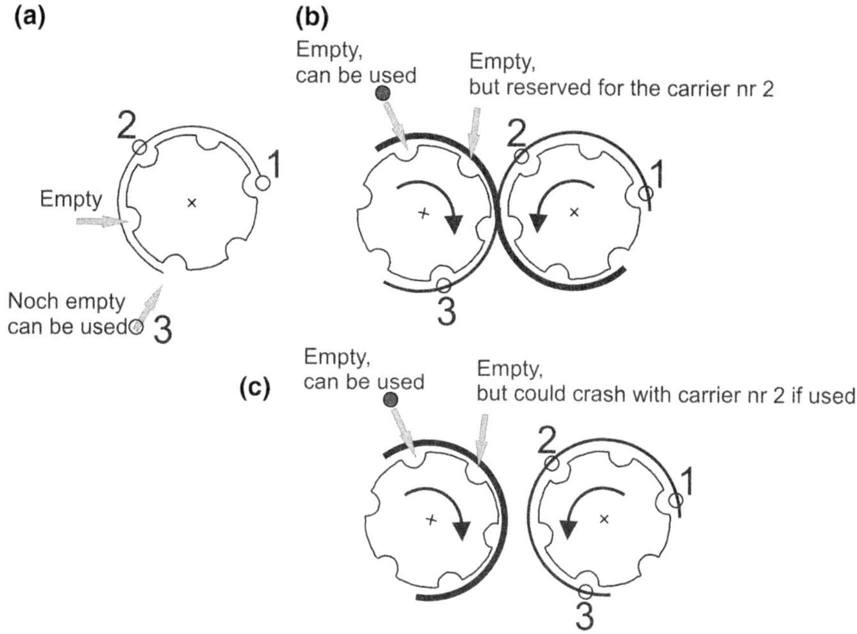

Fig. 4.17 Checking the slot occupation during the carrier arrangement **a** for the first track the condition is for the slot to be empty, **b** for every further track and its places, which crosses itself again, it has to be checked additionally, if the empty places will be required from the already placed carriers, **c** checking the slot occupation of two tracks, which do not really work together, but share a common space

the slots according to a scheme as for instance "1 Full : 1 Empty". For the track based models the realization of this is simple, as it just goes through the list or the tracks. For the horn gear based models, the algorithm is more complex, but the principle is the same.

Filling the carriers according to a "1 Full : 1 Empty" rule, works as following Fig. 4.17a: the starting slot of the starting horn gear receives carrier number 1. The next slot on this track is kept empty, the next one gets the next carrier etc.

The problems are usually starting after the first track of the tubular braiding machines is filled or after half of the track of a flat braiding machine is filled, because placing carriers on the next requested position can lead to collision at some configurations. The following three conditions have to be satisfied, in order to the carrier arrangement to be free from collision:

- the slot, in which a carrier has to be placed has to be empty (only one carrier per slot is allowed) (Fig. 4.17a);
- the slot, in which a carrier has to be placed, should not be used by carriers from another track at a later point (Fig. 4.17b).

- the slot, in which carrier has to be placed, should have enough free space to take the entire carrier without collision with other machine parts or moving carriers (Fig. 4.17c).

The first check is simple—when the slots do not hold any carriers, their virtual places for the carriers are empty. The second test is more complicated—it requires a full simulation cycle of one repeat of the machine in order to proof, if some of the slots will be used by the other horn gears or not.

For numerically checking the satisfaction of the third condition, again, the entire carrier motion simulation should be started and for every simulation step the *distance* between one carrier and the other carriers has to be checked. This is the most computational intensive test, but for up to one hundred carriers this is not problematic using modern computers and by doing it this way, checking the distance keeps the entire algorithm simple and applicable for every horn gear configuration.

4.6.1 Motion Emulation and Build of Simplified Virtual Braid

Once the carriers are arranged properly, the motion emulation is simple—it consists of storing the current carrier positions in the form of (x, y, z) lists in separated objects—yarn axes. After all horn gears are rotated for one time step the carriers receive new coordinates and the z position is shifted on a z-direction at the length at which the braid will be taken for one step:

$$\Delta z = V_{Take\_off} \cdot \Delta t \tag{4.15}$$

Normally, during the simulation, the horn gears are set with their original sizes with a diameter for instance of 80 or 180 mm. Because the motion emulation does not simulate the yarn interaction process, but only the carrier motion, the braid has to be scaled, in order to have similar geometrical properties like the real one. Important for this scaling, is that the relation between the angular velocity of the horn gears and the take off speed before and after scaling, remains the same. The angular velocity of each horn gear depends on its diameter, so the relation can be written as rate between the take off speed V_{Take_off} and the horn gear diameter $D_{HornGear}$:

$$\lambda = \frac{V_{Take\_off}}{D_{HornGear}} \tag{4.16}$$

The real braid will have a diameter of the ridge D_{Ridge}, which depends on the yarn diameter:

$$\xi = \frac{D_{HornGear}}{D_{Ridge}} \tag{4.17}$$

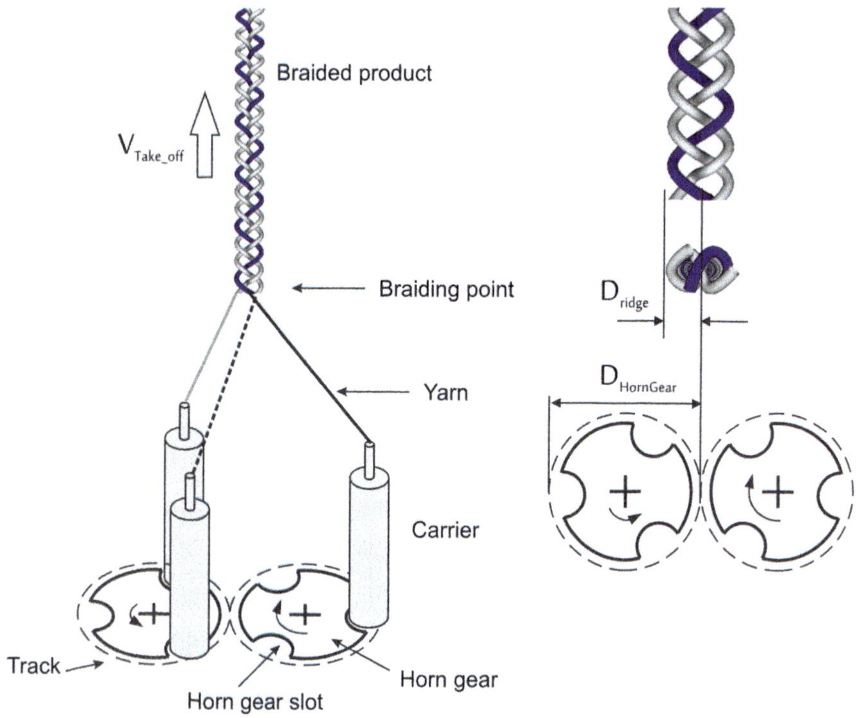

Fig. 4.18 In order the geometry of the braid to be proportional to the reality, the braid or the machine has to be scaled

In order for the 3D image to be realistic, two ways are possible (Fig. 4.18):

- scaling the $x - y$ coordinates so that

$$\begin{aligned} x_{scalled} &= \frac{x}{\xi} \\ y_{scalled} &= \frac{y}{\xi} \end{aligned} \tag{4.18}$$

- or the take of speed and the yarn diameters have to be increased to

$$V_{Take\_off\_scalled} = V_{Take\_off} \cdot \xi \tag{4.19}$$

$$D_{yarn\_scalled} = D_{yarn} \cdot \xi \tag{4.20}$$

The second approach leads to significantly (xi-times) thicker braided structures, which have to be considered, if the simulated geometry should been used for some simulation later.

4.7 Algorithm Implementation

The objects and classes were initially implemented as a track based method in the Matlab language, and then rewritten as discrete track method on C++ as commercial software TexMind Braiding Configurator [16], reported in [11, 13, 18]. For the description of the custom machines a graphical user interface is created under C++ and WxWidgets [22]. The machine data and the carrier arrangement are stored in a specially defined XML based format, which can be used for exchanging the data between different applications. Figure 4.19 presents an example of this format for two horn gears and Fig. 4.20 shows the graphical interface of the current version.

4.8 Simulated Braids

In Fig. 4.21a the configuration of the smallest braiding machine for braiding with three carriers, and three simulated braids with different take off speeds, is given.

Figure 4.22 demonstrates a tubular braiding machine with 16 carriers and a carrier arrangement of “1 Full 1 Empty”, and Fig. 4.23 the same machine with a “2 Full 2 Empty” carrier arrangement. The second configuration demonstrates the empty spaces between the yarns, as described in Chap. 2, which is one of the disadvantages of the method, but from topological point of view, the geometry is correct.

The advantage of using the software starts with the investigation of complex structures as the one, demonstrated in Figs. 4.24 and 4.25. For such a machine, the simplified rules for checking of the carriers on tubular and flat machines are not applicable manually. The generated geometry was the result of an investigation [13] and results of about two days of trials of about 12 configurations by a person with knowledge about braiding, but without experience in the configuration of machines. The virtual tests have shown, that the initial configuration Fig. 4.24a at an arrangement of “1 Full 1 Empty” leads to carrier collision, but the arrangement “1 Full 3 Empty” produces a suitable structure. Actually, using the “1 Full 3 Empty” structure is connected with the usage of less carriers or less yarns in the braid, so it will be more loose. Changing the horn gears at the crossing gears between the single flat parts to such with 5 slots allows a more dense structure, which is shown in Fig. 4.25.

```
<Machine Stitch="80" SlotLength="62,8319" Nr_carriers="0" Nr_horngears="2">
    <HornGears>
        <HornGear-0 ID="0" NumberSections="3" Radius="30" CenterX="0"
        CenterY="100" MovingDirection="0" StartAngle="0" />
        <HornGear-1 ID="1" NumberSections="4" Radius="40" CenterX="100"
        CenterY="140" MovingDirection="0" StartAngle="0" />
    </HornGears>
    <Tracks />
</Machine>
```

Fig. 4.19 Example of XML code for description of a custom machine

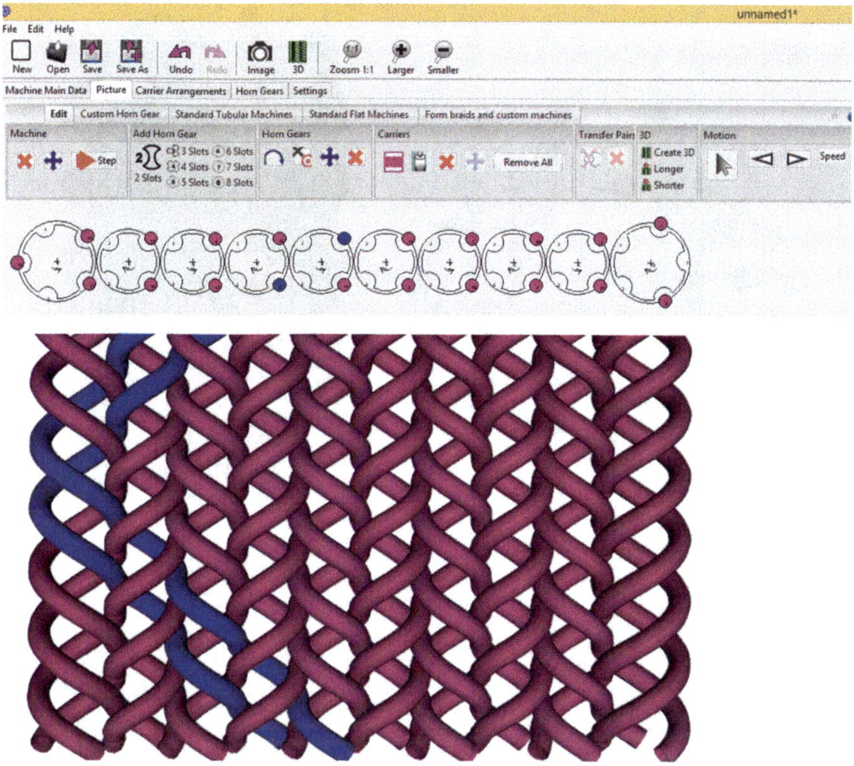

Fig. 4.20 Graphical user interface for configuration of custom machine

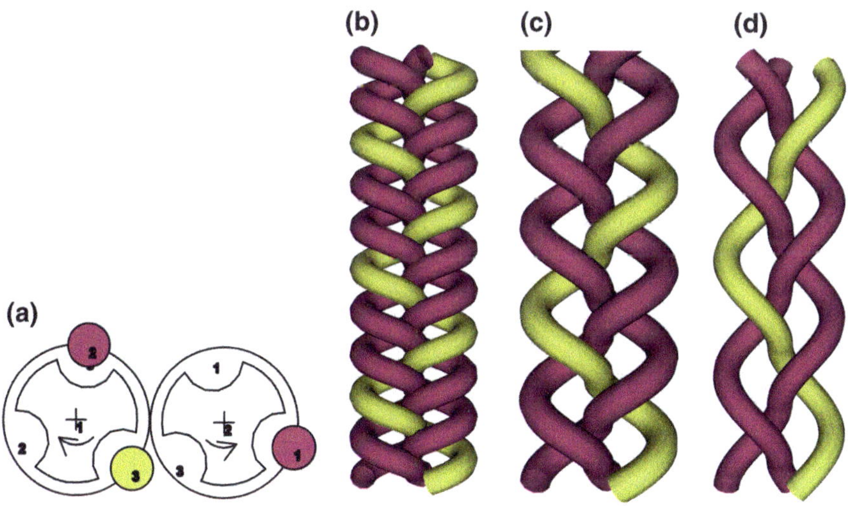

Fig. 4.21 Machine configuration of the smallest braid—with 3 yarns (**a**) and the simulated braids using three different take off speeds (**b**, **c** and **d**), which leads to different braiding angle

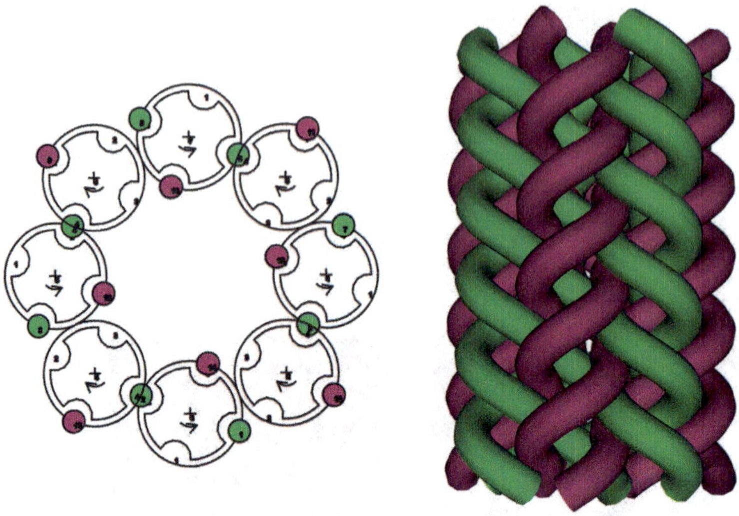

Fig. 4.22 Configuration of tubular braiding machine with carrier arrangement 1 Full 1 Empty and the simulated braid

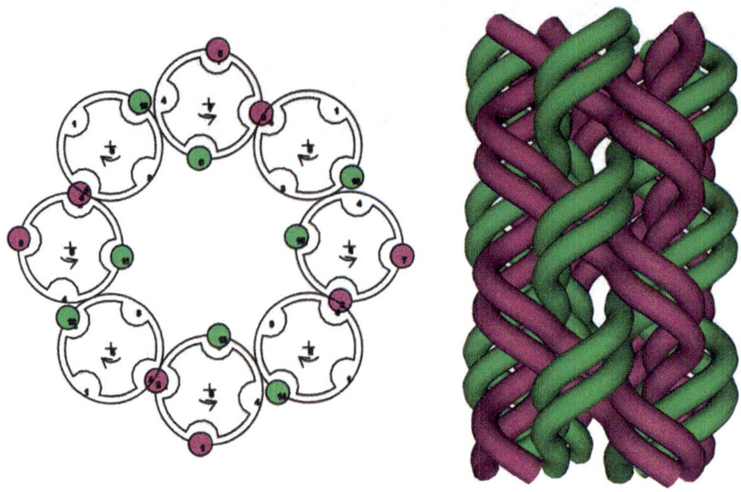

Fig. 4.23 Configuration of tubular braiding machine with carrier arrangement 2 Full 2 Empty and the simulated braid. For this configuration the open spaces from the larger empty space are well visible

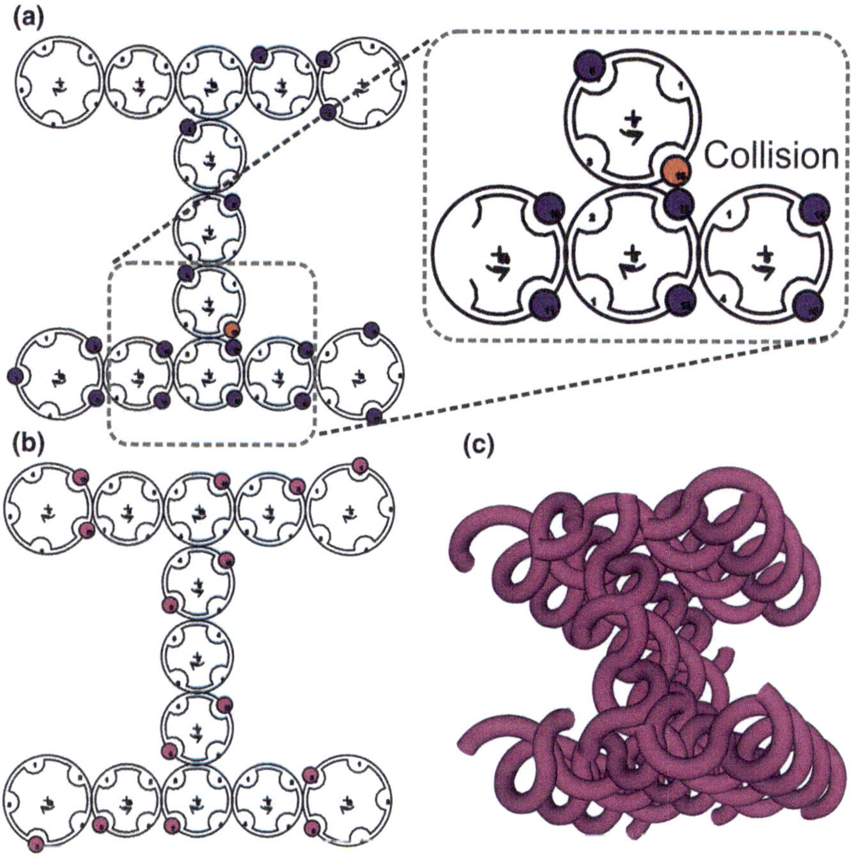

Fig. 4.24 Possible configuration of machine for production of double-T profile and the simulated structure [12]

4.9 Future Steps

All of the algorithms can be connected to a FEM or another kind (mass-point, resp. "digital-chain") of software for the simulation of a complete braiding process. This way, they allow not only the configuration of the machine, but also the numerical investigation and optimization of the parameters as position of the former rings, yarn tension, orientation of the mandrel or core etc. which will additionally speed up the development of new products.

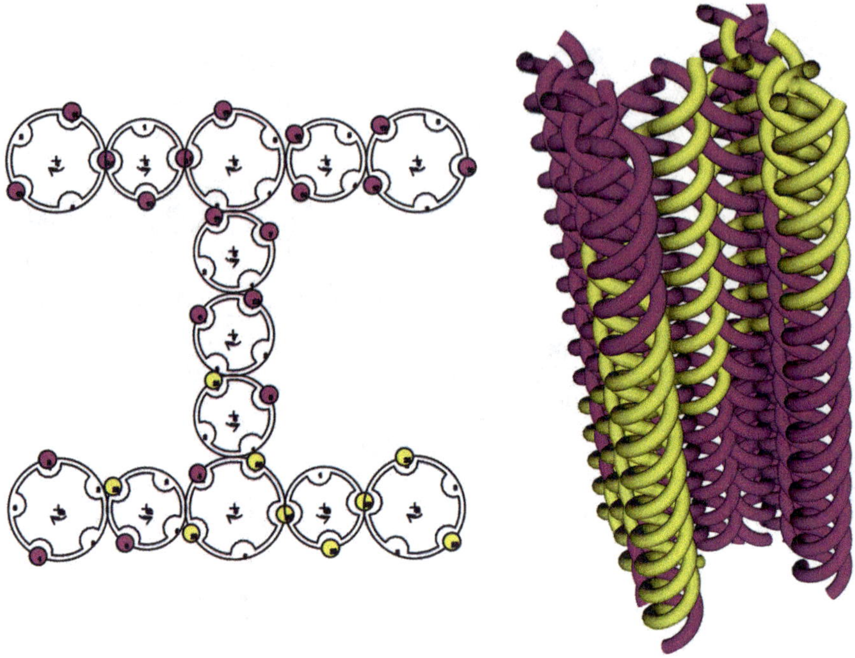

Fig. 4.25 Improved configuration of machine for production of double-T profile and the simulated structure

4.10 Conclusions

Presented are two algorithms—horn gear based and track based—in two variants, which allow the effective emulation of the braiding process of any maypole braiding machine. These are especially useful for developing machines for the production of profiles with various cross sections, for which the horn gear configuration and the carrier occupation are complex and can not be checked manually. Each of the algorithms can be used as a basis for the complete braiding process simulation, for which the carrier motion is required.

References

1. Akiyama, Y., Hamada, H., Maekawa, Z., Uratani, Y., Yokoyama, A.: Braid and braiding method (1995). http://www.google.es/patents/US5385077
2. Akkerman, R., Villa, R.B.H.: Braiding simulation for rtm preforms. In: Long, A. (ed.) TexComp8: 8th International Conference on Textile Composites. Nottingham, UK (2006)
3. Bilisik, K.: Three-dimensional braiding for composites: a review. Text. Res. J. **83**(13), 1414–1436 (2013). https://doi.org/10.1177/0040517512450766

4. Bilisik, K.: Cartesian 3D braiding. In: Kyosev, Y. (ed.) Advances in Braiding Technology. Woodhead Publishing in Textiles, vol. 177, pp. 107–145. Woodhead Publishing (2016). https://doi.org/10.1016/B978-0-08-100407-4.00004-1
5. Bogdanovich, A.E.: An overview of three-dimensional braiding technologies. In: Kyosev, Y. (ed.) Advances in Braiding Technology. Woodhead Publishing in Textiles, vol. 177, pp. 3–78. Woodhead Publishing (2016). https://doi.org/10.1016/B978-0-08-100407-4.00001-6
6. Bradford, P.D., Bogdanovich, A.E.: Electrical conductivity study of carbon nanotube yarns, 3-D hybrid braids and their composites. J. Compos. Mater. **42**(15), 1533–1545 (2008)
7. Brookstein, D., Dent, R., Dent, J., Rose, D., Skelton, J.: Solid braid structure (1994). https://www.google.es/patents/US5357839
8. Büsgen, A.: Neue Verfahren zur Herstellung von dreidimensionalen Textilien für den Einsatz in Faserverbundwerkstoffen: Dissertation (1993)
9. Kahan, W.: Pracniques: further remarks on reducing truncation errors. Commun. ACM **8**(1), 40 (1965)
10. Kessels, J., Akkerman, R.: Prediction of the yarn trajectories on complex braided preforms. Compos. Part A: Appl. Sci. Manuf. **33**(8), 1073–1081 (2002). https://doi.org/10.1016/S1359-835X(02)00075-1
11. Küster, K., Aurich, M., Kyosev, Y.: Machine configurations for multilayer braided structures with complex cross sections. In: Avadanei, M. (ed.) 16th Romanian Textiles and Leather Conference CORTEP 2016, pp. 244–249 (2016)
12. Kyosev, Y.: Simulation driven design of braided products. In: Mahltig, B. (ed.) Textiles: Advances in Research and Applications. Nova Science Publishers, Inc., Hauppauge NY, USA (2017) (in print)
13. Kyosev, Y., Küster, K.: Development and modelling of braided products with complex cross sections. In: Boussu, F., Chen, X. (eds.) Proceedings of the 7th World Conference 3D Fabrics and Their Applications, pp. 209–216 (2016)
14. Kyosev, Y.: Braiding Technology for Textiles: Principles, Design and Processes. Woodhead Publishing Series in Textiles, No. 158, 1 edn. Woodhead Publishing Limited (2014)
15. Kyosev, Y.: Machine configurator for braided composite profiles with arbitrary cross section. In: 16th European Conference on Composite Materials ECCM 16, Seville, Spain (2014)
16. Kyosev, Y.: Texmind braiding machine configurator (2016). http://configurator.texmind.com
17. Lengersdorf, M., Gries, T.: Three-dimensional (3D)-maypole braiding. In: Kyosev, Y. (ed.) Advances in Braiding Technology. Woodhead Publishing in Textiles, vol. 177, pp. 89–105. Woodhead Publishing (2016). https://doi.org/10.1016/B978-0-08-100407-4.00003-X
18. Li, M.: Braiding for composites—investigation about the braiding of products with non-tubular cross section. Master's thesis, Hochschule Niederrhein, Mönchengladbach (2013)
19. Pickett, A.K., Sirtautas, J., Erber, A.: Braiding simulation and prediction of mechanical properties. Appl. Compos. Mater. **16**(6), 345–364 (2009). https://doi.org/10.1007/s10443-009-9102-x
20. Schneider, M.: Konstruktion von dreidimensional geflochtenen Verstärkungstextilien für Faserverbundwerkstoffe. Berichte aus der Werkstofftechnik. Shaker Verlag, Aachen (2000)
21. Schreiber, F.: Three-dimensional hexagonal braiding. In: Kyosev, Y. (ed.) Advances in Braiding Technology. Woodhead Publishing in Textiles, vol. 177, pp. 79–88. Woodhead Publishing (2016). https://doi.org/10.1016/B978-0-08-100407-4.00002-8
22. Smart, J., Roebling R., Zeitlin, V., et al.: wxwidgets cross platform gui library (2015). http://www.wxwidgets.org/
23. Temple, S.: Woven tubular structure (1984). https://www.google.es/patents/EP0113196A1?cl=en
24. Uozomi T., Iwahori Y., Iwasaw S., et al.: Braiding technologies for airplane applications using RTM process. In: Proceedings of the seventh Japan International SAMPE Symposium (2001)
25. van Ravenhorst, J.H., Akkerman, R.: A yarn interaction model for circular braiding. Compos. Part A: Appl. Sci. Manuf. **81**, 254–263 (2016). https://doi.org/10.1016/j.compositesa.2015.11.026
26. Wolfe, J.M.: Reducing truncation errors by programming. Commun. ACM **7**(June), 355–356 (1964)

Part III
Knitted Structures

Chapter 5
Topological Modelling of Knitted Structures

5.1 Classification of Knitted Structures

There are different classifications of knitted structures depending on the classification criteria [15]. Because the task of this chapter is the 3D modelling of these structures, only the criteria, which are relevant for the structural elements, are applied. More details and other classifications can be found in [26, 34, 37, 38].

There are two basic types of knitted structures that depend on the direction of creation of the loops—weft knitted and warp knitted (Fig. 5.1). The most simple way to recognize the fabric is to follow the way of the loop foots. If one foot is coming from the bottom and the other goes up (F1) the structure is warp knitted, and if the foots are moving from left to right (F2)—weft knitted.

A weft knitted structure is usually created from a single yarn or a few yarns, building all the loops in one row (horizontally). A warp knitted structure uses one or more *warp* beams with at least as many yarns as number of needles and all these yarns are building only one loop/needle in one row (with some exceptions for some special patterns) (vertically).

Furthermore, depending on the number of the working needle beds, both, the warp and weft knitted structures, can be divided into two groups:

- single-faced (produced usually on a single-needle bed machine) and
- double-faced (produced on a double-needle bed machine).

The single-faced fabrics are made from loops, which are all faced in the same direction. The double-faced fabrics have two sets of loops so that one set of loops shows the faces to one side and the other set of loops shows the faces to the other side. In Fig. 5.2a, b the face side of a loop is presented and in Fig. 5.2c, d the reverse side is shown. More details about the structures and classification can be found in [34].

For the weft knitting technique (fabrics), there are additional classes of structures:

- purl—where face and reverse stitches are on the same side in the same wale (using double headed needles or transfer technology),

Y. Kyosev, *Topology-Based Modeling of Textile Structures and Their Joint Assemblies*, https://doi.org/10.1007/978-3-030-02541-0_5

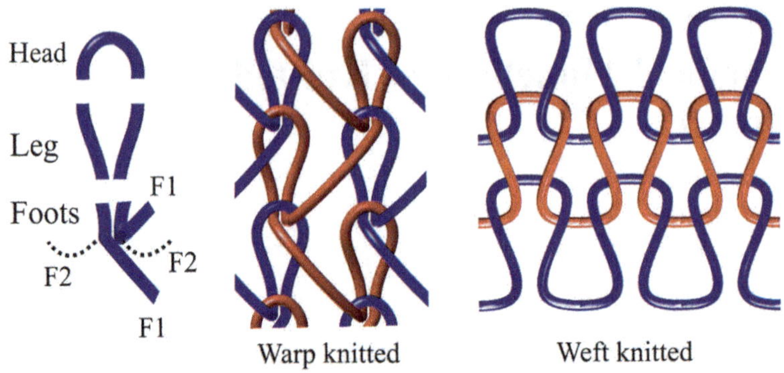

Fig. 5.1 Loop elements and the main types of knitted structures. Reprinted from X. Chen (Ed.) "Modelling and predicting textile behaviour", Y. Kyosev, W. Renkens, "Modelling and visualization of knitted fabrics", pp. 225–262, Copyright (2010) with permission from Elsevier [15]

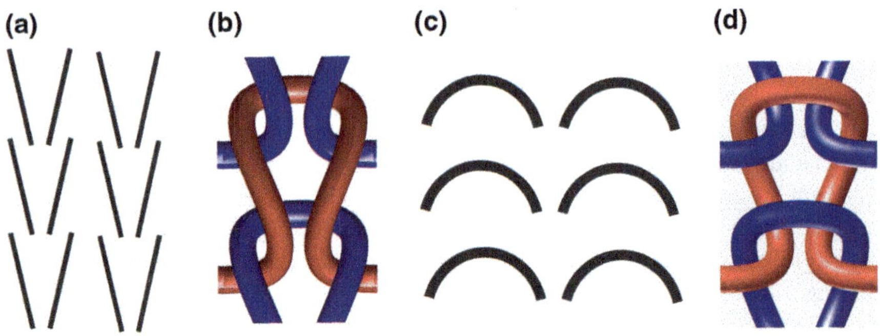

Fig. 5.2 Loop elements and the main types of knitted structures

- interlock—could be classified as a special (sub-) class of double-faced structures, because the loops on both sets are not by half a needle step, like the ones of double face structures. But from a modelling point of view, these structures can be treated as derivatives of the double bed structures. They can be reconstructed using the same loops, but using a special arrangement for their position and orientation.

Single-faced warp knitted structures can be divided into subclasses, depending on the number and arrangement of yarns used during the knitting, like:

- single guide bar with full threading,
- multiple guide bars with full threading and
- multiple guide bars with partial threading (general case) [26].

Depending on the number of the guide bars and the threading, the number of yarns building the loops (or other elements) on the same needle during a cycle, can be calculated. This information is important for the modelling. With this data the general orientation of the yarns in the loop can be calculated. Analyzing and modelling of some particular sub-class structures does not make sense—all the subclasses are presented in the double-faced warp knitted structures and because of this, they all have to be modelled by one general method.

For the modelling of knitted structures, another criterion for the classification could be very useful: the types of different structural elements in a fabric. Fabrics can be divided into fabrics having only plain loops or loops combined with one or more of the elements—tucks, plating loops, transferred loops, weft insertions etc. Since each of these structural elements has to be modelled separately, they will be discussed in more detail in the next section.

5.2 Scales in the Structure

Knitted structures consist of a large set of structural elements, described in the next section. These structural elements are built from yarns, which are presented as single curved tubes for simplification reasons. In reality, these yarns are structures of single fibres or filaments.

The three levels—structure, structural elements, and single fibres, present three scales named macro-, meso- and micro-scale levels [17, 36].

The macro-scale is the most important one for the applications. It handles the mechanical behaviour of the structure as a continuum membrane or plate with known properties (Fig. 5.3). There are different methods to obtain the information at this macro-scale—the most used way is to make different mechanical tests with the samples. In order to understand the macro-scale behaviour or to predict it for some cases, the behaviour at the meso-scale is required. At the meso-scale, the yarns are considered as continuum with known properties. At this level, the geometry and the mechanics of the single unit cell of the structure can be considered as known and conclusions can be drawn from the behaviour of the materials. Meso-scale models neglect the effects at the micro-scale level, which is the result of single fibre-to-fibre interactions.

To model the entire knitted structure successfully, all these levels have to be considered in some matter during the modelling process. The accuracy of the model depends on level of preciseness at which the modelling is done and on the fact, for which levels only the main information is used. For example—if after some investigation, the exact yarn cross section form is not important, then the micro level—the interactions between the fibers inside the yarn—is not required. In other cases, such as medical filters, the distance and the distribution of single fibres could be of higher importance. Then the micro-level has to be modelled and connected with the meso-level into a multi-scale model.

In which way the information at the different levels will be modelled (or in which way the information of the different levels will be modelled), does not play a significant role for the model. The higher the required accuracy of the models, the more details of the levels have to be considered.

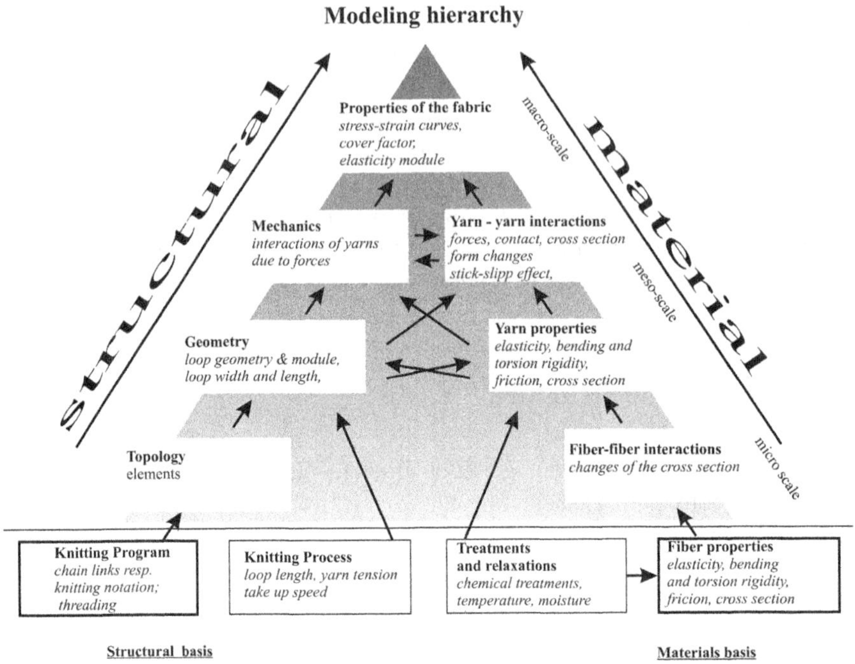

Fig. 5.3 Scales in the knitting modelling. Reprinted from X. Chen (Ed.) "Modelling and predicting textile behaviour", Y. Kyosev, W. Renkens, "Modelling and visualization of knitted fabrics", pp. 225–262, Copyright (2010) with permission from Elsevier [15]

There are two basic sources of information (Fig. 5.3), which determine most parameters of knitted structure: the knitting program, as topological basis, and the fiber properties, as material basis. With this information, the typical behaviour of the knitted structure can already be predicted. The knitting process parameters and the treatment of the structure are presenting secondary information, which makes the model more accurate and more exhaustive in geometry and dimensions.

There are different ways of building models, depending on how the interactions between the levels are considered. Including all the levels from micro up to macro in one model is not effective until now. The high complexity of the nature and the large number of connections between the parameters make such models very complicated and difficult to handle. Usually, only one of the levels is modelled and the information from the lower levels is given in some condensed form. For the knitted structures the most important in this case is the yarn level, as it defines the flexibility of the structures and its main mechanical parameters.

5.3 Structural Elements of Knitted Structures at Meso-scale

Woven and braided fabrics can be represented as a set of single structural elements—crossing yarn pieces. Depending on which yarn (weft or warp) is in the upper position, these elements can be coded, creating the weaving pattern. This approach was extended for multilayer fabrics, where instead of warp or "not warp" a number, corresponding to the layer, was used [2, 19]. The attempt to use this approach for knitted structures was implemented in software for weft knitted structures [33]. The program can produce 2D pictures, but requires a special way of coding—since two cross points have to be given for a single loop. This approach requires that the user has special knowledge about the position of the yarns and thus did not find wide usage in modelling.

Most authors use the whole loop as a single structural element within the modelling. In one loop, several single cross point pairs are included, but they appear at every loop, and therefore it is quite effective to make this kind of coding at the loop level. This way of coding the pattern is widely used in commercial software like M1 of Fa. H. Stoll GmbH & Co. KG, Reutlingen, Germany, as well as in the academic community like WeftKnit [7, 22, 23] etc. Additionally to the main element—the loop—different structural elements can be found in knitted structures, presented in Fig. 5.4—floatings, built at places of missed stitches (a) or as underlapping (b); tucks in their real form (d) or idealized with longer loop before the tuck (d); transferred loops (f) with its new build element N consisting of legs and head only; partial weft (g) or full weft (not presented, going through the complete structure); plated stitches (h) and (i) where one loop is based build of two yarns. For each of these elements, separated geometrical descriptions have to be done, if these are available in the structure, which has to be modelled.

5.4 Topology Description

The graphical description of the knitting process (and its program) is a simplified way to notate the topology of knitted structures. It has the information to understand the topology of the structure, but it does not present the yarn interconnections exactly in the way, as it is already available in the 2D representation (Fig. 5.5).

For cases, which do not require an exact 3D geometry of the knitted structure, but just the optical appearance of this structure, it is suitable to use the notations as a pattern and to map them to receive the elements pictures of the corresponded elements [20, 21]. For normal engineering applications, which require the real 3D representation, a set of points of the loops has to be created. A logical way to do this, is to start from the 2D model and then add the third dimension.

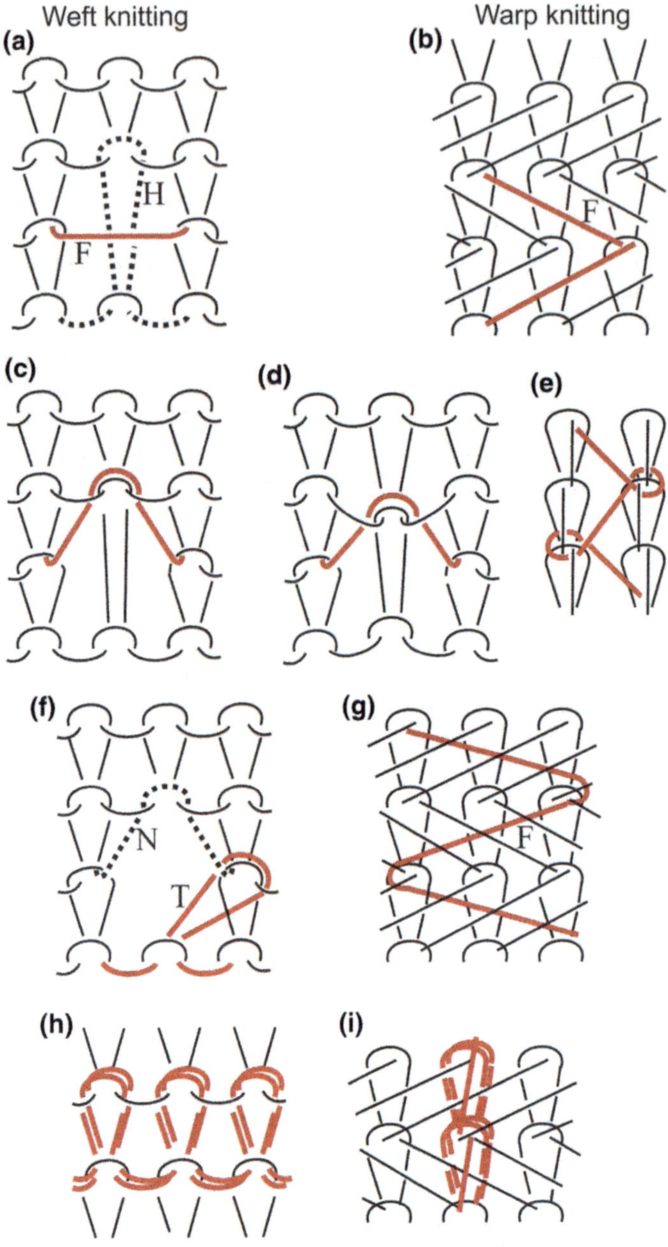

Fig. 5.4 Structural elements in the knitted structures

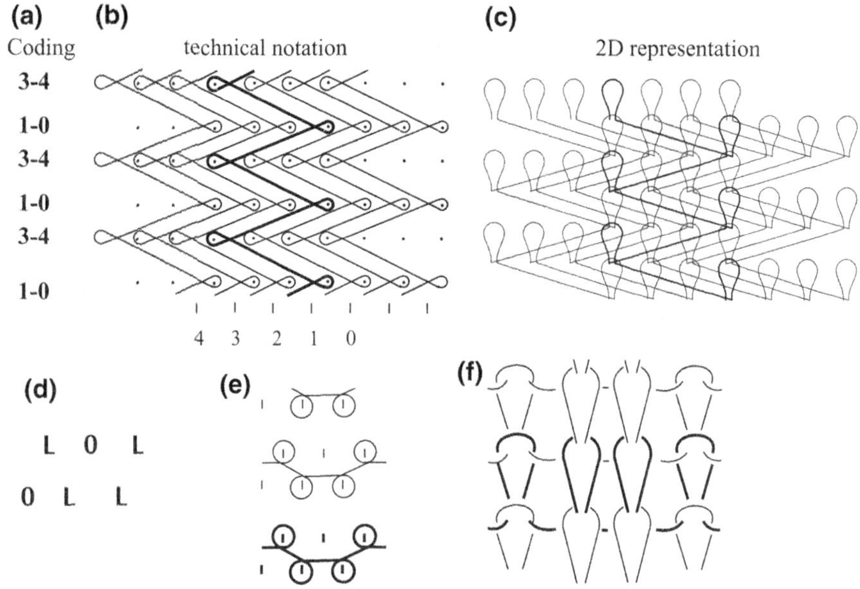

Fig. 5.5 Numerical coding techniques (**a**) and (**d**), technical drawing (**b**) and (**e**) and 2D representation (**c**) and (**f**) of warp and weft knitted fabrics, respectively. Reprinted from X. Chen (Ed.) "Modelling and predicting textile behaviour", Y. Kyosev, W. Renkens, "Modelling and visualization of knitted fabrics", pp. 225–262, Copyright (2010) with permission from Elsevier [15]

5.5 Loops and Loops Only Based Knitted Structures

Following the definition of Spencer [34] "the interlooping consists of forming yarn(s) into loops, each of which is typically only released after a succeeding loop has been formed and intermeshed with it so that a secure ground loop structure is achieved. The loops are also held together by the yarn passing from one to the next". This rule is valid for all structures, held together by loops—weft knitted, warp knitted and multilayer structures, in which thin loop yarn holds together the heavy tows.

A loop consists of a head H (Fig. 5.1), two side legs (limbs) L, at each base is a foot (F). The foot meshes through the head of the loop, which was formed at the previous knitting cycle, usually by the same needle. The yarn passes from the foot of one loop into the foot and leg of the next loop formed by it. In weft knitting the yarn (normally) passes to the left or right in the *same* row, while in warp knitting one foot passes to the next loop in the *following* row. If the incoming and outgoing foots are crossing, the loop is referred to as closed (Fig. 5.6c), if they are not crossing it is called open loop (Fig. 5.6b). The subdivision into open and closed loops is important only for warp knitted structures. In weft knitted structures the feet are normally open, because the yarn continues to be supplied in one direction Fig. 5.6a. Some exceptions—the closed weft knitted loops can be found, but these are mainly knitted manually or on specially designed machines for this.

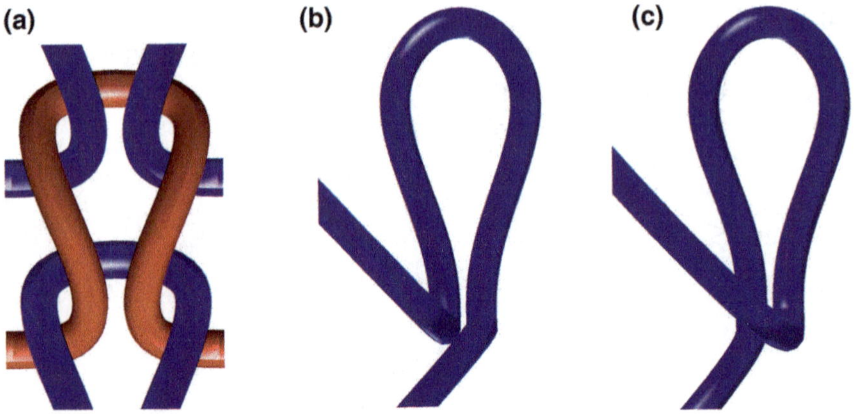

Fig. 5.6 Depending on the direction of the foots, the warp knitted loops can be open (**a** and **b**) or closed (**c**)

5.5.1 Early Analytical Models for Geometry Relations of Knitted Loops

The engineers, in contrast to the physicians and mathematicians, are looking for a simplified solution to the problems based on the knowledge of the process. A simulation of the knitting process is not required in order to obtain the maximal elongation of the yarns, during this knitting process. There are some simple calculations to check, if the production of the fabrics is possible.

The standard check for knitted fabrics begins with the compatibility of the yarn thickness and the available space in the machine and structure for the yarn. For both, warp and weft knitting, the yarn generally may not be thicker than 1/4 of the distance between the needles.

From the well known relations of the loop geometry, developed by Doyle and Munden [4, 25, 27]:

$$\begin{aligned} c &= \tfrac{k_c}{l} \\ w &= \tfrac{k_w}{l} \\ \tfrac{c}{w} &= \tfrac{k_c}{k_w} = 1.3 \end{aligned} \tag{5.1}$$

and for the tightness factor of the fabric K

$$K = \frac{\sqrt{tex}}{l} \tag{5.2}$$

it can be checked, if the planned fabrics belong to the "normal" cases or, if they can be considered as special cases. In the equations presented above, c is the number of courses per unit length, w is the number of wales per unit lengths, k_c and k_w are constants, tex is the yarn linear density (g/km), and l is the loop length in (cm).

According to [27], the constants in normal cases are in the ranges of $k_c = 5.0 \div 5.5$, $k_w = 3.8 \div 4.1$, $K = 13 \div 15$. More detailed statistical data about these coefficients for weft knitted fabrics can be found in [9].

The next step is done to verify, if all loops are interlooped. If this is not the case, an open, non stable structure will be build. A possible way for the implementation of such rules is well described in [21]. At this point, it has to be mentioned, that there are some special knitting techniques, within which the single yarns are overlapping on the needle, but are not knitted with the next loop afterwards. This way, free yarn length for the neighbour loops is reserved, which is important for the reduction of the yarn tension during the knitting. For instance, the "braid" pattern has several loops that have to be transferred to a needle and then moved to one or the other side in several positions. Similar cases require special attention, which is not possible in this work. But such simple rules and checks used together with physical simulation or expert system, allow a quick check of the input data. It makes the developing process of the algorithms more convenient and protects the users of the modelling software from logical errors and mistakes.

5.5.2 *Analytical Model of Choi*

A very simple method, using parametric description of the loop central axis, is presented in [3]. It is based on the assumption, that the yarn cross section is circular and in the very well selected functions for interpolation of the loop form:

$$
\begin{aligned}
y(t) &= 0.5\,(c + 2e)\,(1 - \cos \pi t) \\
x(t) &= at^3 - 1.5at^2 + 0.5\,(a + w)\,t \\
z(t) &= 0.5\,(t_h - d)\,(1 - \cos 2\pi t)
\end{aligned} \tag{5.3}
$$

Equations 5.3 model half of the loop, with a space curve parameter of $t = 0 \div 1$. Here a is a free parameter, c, w respectively describe courses and wales, t_h is the fabric thickness, d is the yarn diameter and e is the adjacent overlapping distance. By given course and wale dimensions and yarn diameter, the rest of the parameters are obtained from geometrical dependencies between loop dimensions and positions of the characteristic points.

5.5.3 *Extended Analytical Model for Elliptical Yarn Cross Section*

Most of the knitted structures are produced by multifilament yarns. These yarns are being bent to a curvature radius approximately equal to the yarn radius. By such bending, the cross section of the bent part will not remain circular, and will be deformed. In the areas with small radius of curvature, the thickness becomes smaller, which leads to the hypothesis of an oval (ellipse-like) form of the cross-section. Researchers of yarn cross-sections use and demonstrate this form for cases of lateral loading of the yarns. In [5, 6] the theory of the yarn cross section for a two-component yarn, based on the density of fibre distributions, is presented. This theory uses two yarn diameters for the characterization of the cross section—the first in the direction of the load and the second in the perpendicular direction. This idea is applied in WiseTex software [17] and the oval form of the yarn is widely used in pre-processing and some calculations of woven and braided fabrics.

In knitted structures the yarns can be highly compressed by two sides or negligibly compressed, depending on the fabric density. The high curvature of the yarn axis pre-conditions the initial yarn cross section to be also changed to one with a higher density of the fibre distribution. Let three neighbour fibres or monofilaments be bent with curvature radius close to the yarn radius.

Their positions after bending could be presented as seen in Fig. 5.7a, as this configuration remains their length constant. The configuration in Fig. 5.7b needs the

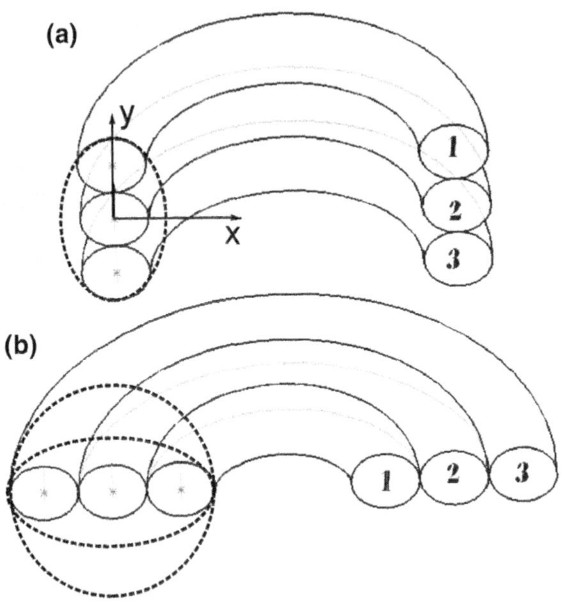

Fig. 5.7 Orientation of single filaments in multiply yarns during bending and approximation of the cross section with elliptic form **a** all yarns cross sections are distributed along the bending axis—the system has minimum of bending energy and all yarns have the same length; **b** worst case of orientation, which can exist in the nature only with application of additional constraints. Reprinted with permission from Yordan Kyosev, Y. Angelova, R. Kovar, (2005) "3D Modeling of Plain Weft Knitted Structures of Compressible Yarn", Research Journal of Textile and Apparel, Vol. 9 Issue: 1, pp. 88–97, https://doi.org/10.1108/RJTA-09-01-2005-B009 (C) Emerald Publishing Limited all rights reserved [14]

outer fibres to be extended and the inner ones—compressed, which demands more energy compared to a small change of the position. The orientation of the smaller diameter of the oval in the direction of the normal to yarn axis, and the greater—in the binormal could be proven also by the minimization of the total strain energy of the yarn. Following [27], the total energy of the knitted structure could be presented as the sum of energy terms of each individual loop:

$$E = \sum_i \int_0^{L_i} (E_b + E_r + E_c + E_t)_i ds \tag{5.4}$$

where E_b, E_r, E_c and E_t are, respectively, the energy terms per unit length of yarn for bending, torsion, lateral compression and longitudinal tension and L_i is the length of any loop. In [27] each of these terms is considered with constant cross section properties as bending rigidity B and torsion rigidity G, although the yarn compression is considered. The changes of the yarn cross section also change the geometrical moment of inertia J of the cross section, which influences the rigidities, as $B = EJ$. As a result, one more accurate minimization of total energy and each of the terms of it cause a problem, not only with the yarn axis form, but also in the yarn cross section change. Assuming that the yarns are linear elastic in bending and the yarn axis has curvature $\kappa(s)$ in the point s, the energy for bending of length L is

$$E_B = \frac{1}{2} \int_{s=0}^{L} B(s) \cdot \kappa(s)^2 ds \tag{5.5}$$

Analogously, for all the terms in Eq. 5.4 cross section parameters can also be considered, but here only bending is analyzed, as it has the greatest term in Eq. 5.4, because of the very high curvature. The stable state of the loop has minimum energy, which will be true, if both B and κ are minimal. This means, that a stable loop configuration depends not only on the curvature (or loop axis form), but also on cross section parameters as moment of inertia, included as a part of the bending rigidity. If the fibres are distributed uniformly in the elliptic cross section, the minimal moment of inertia is one, relative to the larger diameter d_2 (Axis y for Fig. 5.7a)

$$I_y = \frac{\pi}{64} d_1^3 d_2 \tag{5.6}$$

For complex 3D curves, this orientation means that the smaller diameter d_1 of the ellipse remains in the same direction with an axis normal to the current point, and the greater diameter d_1—parallel to the axis binormal. A demonstration of this idea can be found in Fig. 5.7—in the case (a) the cross section normal vector is parallel to the shorter axis of the ellipse, this leads to lower moment of inertia and equal yarn length distribution. Case (b) on the same picture can be produced only with external forces during special forming of the yarns, where different lengths are allowed.

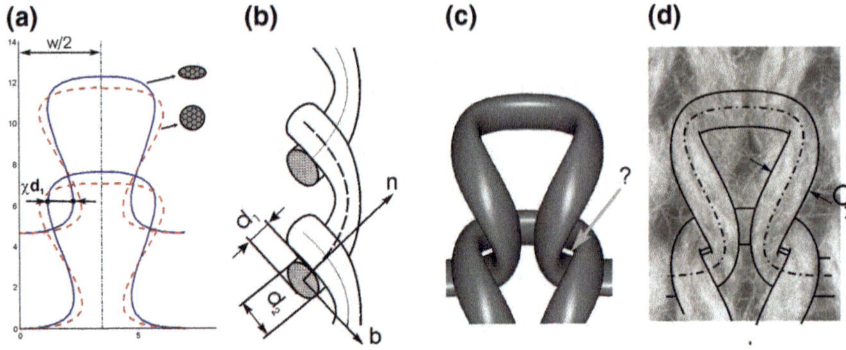

Fig. 5.8 **a** Required distances or forms for circular cross section and for ellyptic cross section, applying the model of Choi [3], **b** orientation of the cross section, **c** resulting empty spaces, **d** comparison with the real structure. Reprinted with permission from Yordan Kyosev, Y. Angelova, R. Kovar, (2005) "3D Modeling of Plain Weft Knitted Structures of Compressible Yarn", Research Journal of Textile and Apparel, Vol. 9 Issue: 1, pp. 88–97, https://doi.org/10.1108/RJTA-09-01-2005-B009 (C) Emerald Publishing Limited all rights reserved [14]

A trial to extend the model, based on Eq. 5.3 of Choi [3], is done by Kyosev et al. [14] by the introduction of a non circular cross section, for instance elliptical. Introducing an elliptical cross section in the model presented in [3] requires the distance between touching points (Fig. 5.8a) to be changed to χd_1, as in this contact point, yarns are contacting by their larger sides. Here $\chi \geq 1$ is the loop separation parameter. After calculating the yarn axis, the elliptical cross section is extruded, following the yarn path. The assumption is, that the smaller ellipse diameter d_1 remains parallel to the normal of the axis, as presented in Fig. 5.8b. The 3D loop is presented in Fig. 5.8c and in Fig. 5.8d over the real sample, used for measuring the loop parameters. In the Fig. 5.8c, d, the main problem of this model is identified—there is a large gap between the previous contact areas of the loops.

Increasing the yarn width leads to interpenetrations (Fig. 5.9b). Partial solution can be the increasing the loop height and translating each loop vertically—Fig. 5.9c. This additional correction of the loop height is made in SolidWorks CAD software, analyzing the mutual position of each couple of loops from different points of view and also using the option "collision detection". As the contact area is a complex 3D problem, it is not possible to determine a well working algorithm, only using 2D projections of yarn axis for the calculation of the proper loop's position. For calculating the proper axis form, the whole loop volume should be considered. This is also demonstrated in Fig. 5.9d, in which the loop intersects itself, because the larger diameter of the ellipse d_2 was never used in the calculations.

The calculation of all contact points and definition of the geometrical condition of the yarn axis and for a greater diameter of cross section d_2 is possible, but it requires solving a more complex mathematical problem. Furthermore, its solution will only be used in the current partial case of a plain weft knitting structure.

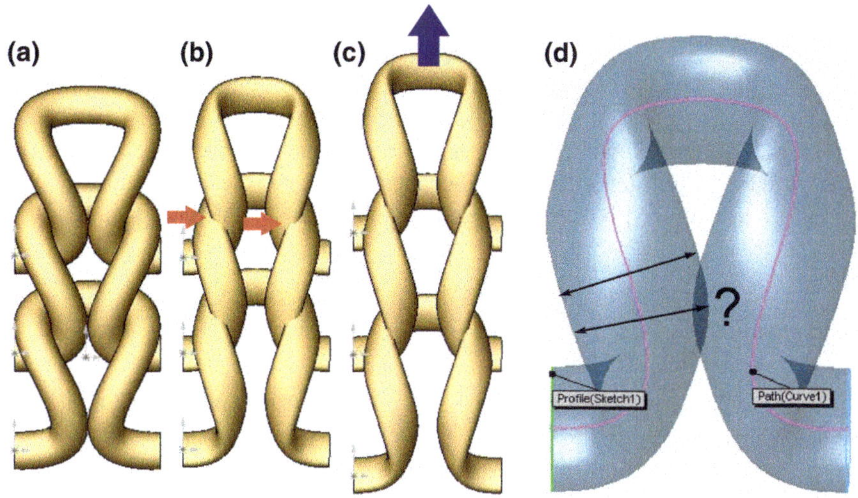

Fig. 5.9 Analytical model with **a** circular cross section, **b** direct application of the elliptic cross section leads to intersections in some areas, **c** trial to adjust the loop positions manually, **d** problem with the self intersection of the loop. Reprinted with permission from Yordan Kyosev, Y. Angelova, R. Kovar, (2005) "3D Modeling of Plain Weft Knitted Structures of Compressible Yarn", Research Journal of Textile and Apparel, Vol. 9 Issue: 1, pp. 88–97, https://doi.org/10.1108/RJTA-09-01-2005-B009 (C) Emerald Publishing Limited all rights reserved [14]

This investigation, reported in [14] demonstrates, that the analytical models can be used for quick calculation of basic forms, but their adjustments or extensions in order to satisfy some requirements in regard to the accuracy or the form of the yarn cross section, are very limited.

Similar models, but based on equations for single parts of the loops were published in the past years by A. Kurbak and his team for different types of weft knitted structures [10, 11] etc. These models have all power and limitations of the analytical models the curves can be implemented quickly and calculated, but the application of each set of equations is limited to the type of the elements for which it is created. As they have less industrial importance, they are explained in less detail within this work.

5.5.4 Topology Using List of Key Points

The topologically correct representation of loops and other structural elements can be defined using a set of key points, connected with some interpolation function. The most commonly used key points are those of the contact between the yarns (or self contact of different yarn pieces of the same yarn). The yarns are contacting often along larger segments. The simplicity of the parametric relations about the minimal distance between the coordinates of the selected key points depends on the choice of the few key points. One of the simplest topological representation is depicted in Fig. 5.10, [23, 24, 39]. The yarn axis of each loop is described as a curve through 6 points, which are the contact points between the loop at the $X - Y$ plane. For this representation, the following parameters of the loops have to be known: the loop high B and the loop width L the course A, the distance between the feet K and the height at this distance yB. With these parameters the points will be defined as follows:

$$\begin{aligned} P.1 &: (x = \pm\tfrac{L}{2};\, y = 0) \\ P.2 &: (x = \pm\tfrac{K}{2};\, y = yB) \end{aligned} \tag{5.7}$$

The coordinates of these points for the loops of the wale i and course j then could be defined as:

$$\begin{aligned} 1_{i,j} &: x = \pm\tfrac{L}{2} + A\cdot i \quad y = 0 + B\cdot j \\ 2_{i,j} &: x = \pm\tfrac{K}{2} + A\cdot i \; y = yB + B\cdot j \end{aligned} \tag{5.8}$$

Using the Eq. 5.8 one loop can be defined as a curve, going through the points

$$1_{1,1} - 2_{1,1} - 1_{2,1} - 1'_{2,1} - 2'_{1,1} - 1'_{1,1} \tag{5.9}$$

The loop height B is defined by the take off speed of the warp knitting machine. For the weft knitting machines, where force driven take up mechanisms are used, B can be defined by the Eq. 5.1. The wale is defined as $A \le \frac{25{,}4}{E}$, where E is the machine gauge. If the fabrics have to be modelled with the same dimensions as they are produced on the machine, then A is equal to the space between the needles. In other cases the structure relaxes and A is usually, but not necessarily smaller.

The factor K and the coordinate yB present the minimal distance between the two yarn axes, and they depend directly on the yarn radius r:

$$\begin{aligned} K &\ge 2\cdot r \\ yB &\ge 2\cdot r \end{aligned} \tag{5.10}$$

At the stage of topological representation, only the location of the yarn axes is important. The form of the yarn cross section does not play any role. If the described equations have to be used for building a more precise geometry of the loop, then the changes of the yarn cross section of staple and multifilament yarns have to be taken into account.

The points $1_{1,1}, 2_{1,1}, 2'_{1,1}$ and $1'_{1,1}$ define the position of a loop head. All coordinates of other loops of a plain structure can be calculated by using a simple translation of these points in the X and Y direction.

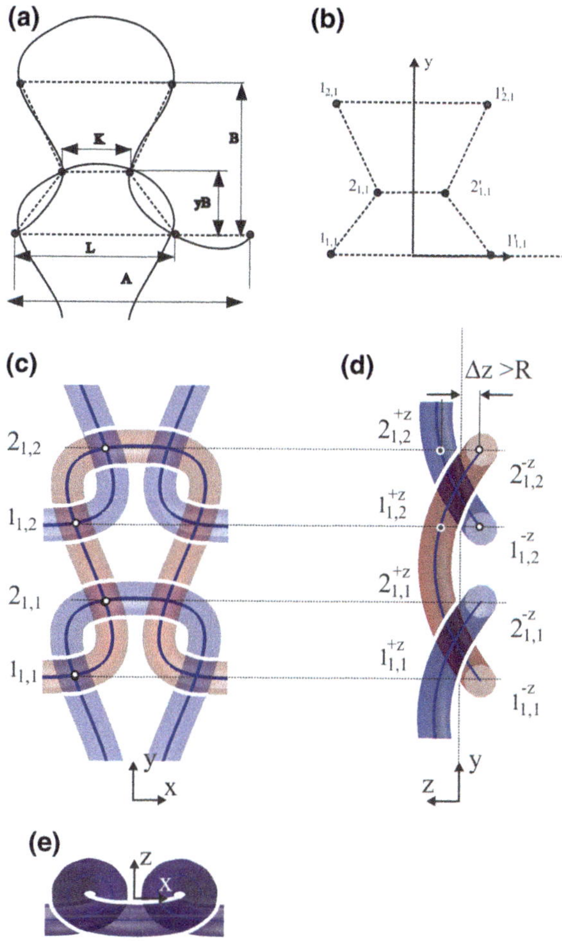

Fig. 5.10 Geometrical models for the 3D representations of loops. Reprinted from X. Chen (Ed.) "Modelling and predicting textile behaviour", Y. Kyosev, W. Renkens, "Modelling and visualization of knitted fabrics", pp. 225–262, Copyright (2010) with permission from Elsevier [15]

The 3D topological representation of the structure allows simplified but real 3D models. It can be derived from the 2D representation, if the coordinates of the points consider the z-axis (Fig. 5.10d)

The point $1_{i,j}$, and the two points $1^{+z}_{i,j}$ and $1^{-z}_{i,j}$ are associated:

$$
\begin{aligned}
1^{+z}_{i,j}(x, y, z) &\equiv \left(1_{i,j}(x), 1_{i,j}(y), +\Delta z\right) \\
1^{-z}_{i,j}(x, y, z) &\equiv \left(1_{i,j}(x), 1_{i,j}(y), -\Delta z\right)
\end{aligned}
\tag{5.11}
$$

where $1_{i,j}(x)$ is the x-coordinate of the point $1_{i,j}$

Here $\Delta z > R$, where R is the yarn radius.

With this notation half of the plain loop can be defined as a curve going through the points: $1^{-z}_{1,1,} - 2^{+z}_{1,1} - 1^{+z}_{1,2} - 2^{-z}_{1,2}$

This principle can be extended and applied to all structural elements. Usually, there is not only one good choice for the position of the key points. Since the goal of this section is only the *topological* representation of the structural elements, the selection of the key-points position is not very important. These points have to be characteristic for the element, they have to be defined easily and the calculation of their coordinates should not be very complex. The loops of the typical warp knitting pattern consist of more then one yarn and therefore more z-positions will be required for their key points. For example—if each loop is built from two yarns, then to the point $1_{i,j}$ the key points $1_{i,j}^{-2z}$, $1_{i,j}^{-z}$, $1_{i,j}^{+z}$, $1_{i,j}^{+2z}$ will be associated.

5.5.5 *Software for Loop Based Structures on One or Two Needle Beds*

Based on the described loop topology, different weft and warp knitted structures can be modelled in the 3D space. For the weft knitted structures, there are different arrangements of face and reverse stitches for the building of purl and interlock structures [34]. These can be built by mirroring the loop head and can be created for instance with the Software WeftKnit [23], KnitGeoModeller [35], using Python script in TexGen [31], TexMind WeftKnitting3D [13], the 3D software of company Shima Seiki [32], presented during ITMA 2015 in Milano, and which is also reported in the works of [8].

The warp knitting structures, based on *loops only*, can be produced on machines working with only one needle bed. The first implementation, based on the equations in industrial software WarpKnit3D of the company ALC Computertechnik GmbH (successor for the part of software Texion GmbH), was done by the author and presented during the ITMA Mnich 2007, as well as in [28]. Figure 5.11 presents such a structure, made from the current implementation of the module Loop3D [16, 29] which integrates well into TexMind Warp Knitting Pattern Editor [13]

5.6 Hold Loops and Tuck Loops

The hold loops and the tucks require points from more wales. That's why the first subscript of the point number shows to which wale (row) it belongs. A hold loop (Fig. 5.12a) then goes through the points $1_{1,1} - 2_{1,1} - 1_{3,1} - 2_{3,1} - 2'_{3,1} - 1'_{3,1} - 2'_{1,1} - 1'_{1,1}$. The tuck loop moves according to Fig. 5.12b through the points $1'_{1,j-1} - 1_{3,j} - 2_{3,j} - 2'_{3,j} - 1'_{3,j} - 1_{1,j+1}$. The real length of the hold loops and the tucks depends on the yarn friction properties, which type and length the neighbour loops have and how the process is controlled in regard to the needle motion. If enough length of the yarn is reserved for these two elements, they can remain as longer loops. Very often, no additional length is reserved for these elements or they are done even shorter,

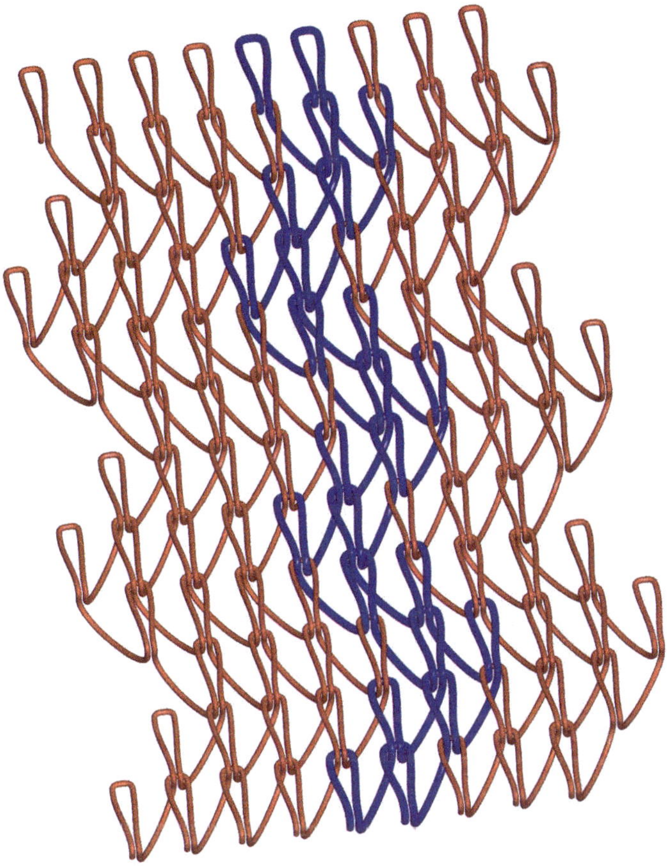

Fig. 5.11 Single needle bar warp knitted fabric, build from loops only

which leads to them pulling down some parts of the structural elements from the next wale, causing the creation of a 3D effects like complex relief of the structure. The consideration of these effects require knowledge of the loop length, before the modelling. For the topologically correct structure, the exact position of the head is not important, as it can be corrected at a later step, using relaxation in mechanical models.

5.7 Plating Loops

Plating is the process of building loops together with two or more yarns (Figs. 5.13, 5.14, 5.15, 5.16, 5.17 and 5.18). For weft knitted structures, plating is used as a special patterning technique—in which one of the yarns usually covers a second one

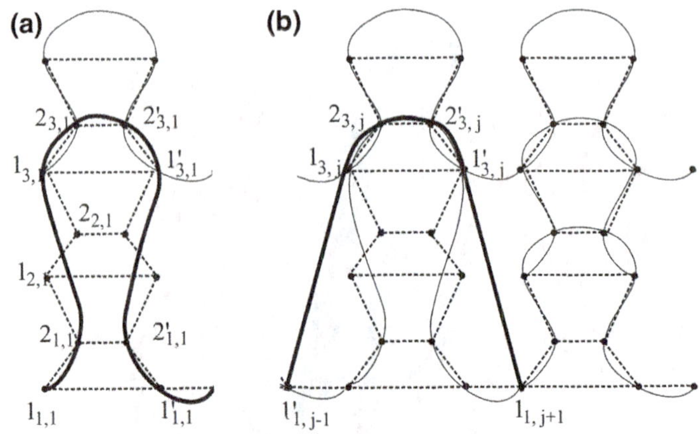

Fig. 5.12 Key points used to define the topology of **a** hold loop and **b** tuck. Both elements require common coordinates with some of the following loops in the vertical direction. Reprinted from X. Chen (Ed.) "Modelling and predicting textile behaviour", Y. Kyosev, W. Renkens, "Modelling and visualization of knitted fabrics", pp. 225–262, Copyright (2010) with permission from Elsevier, [15]

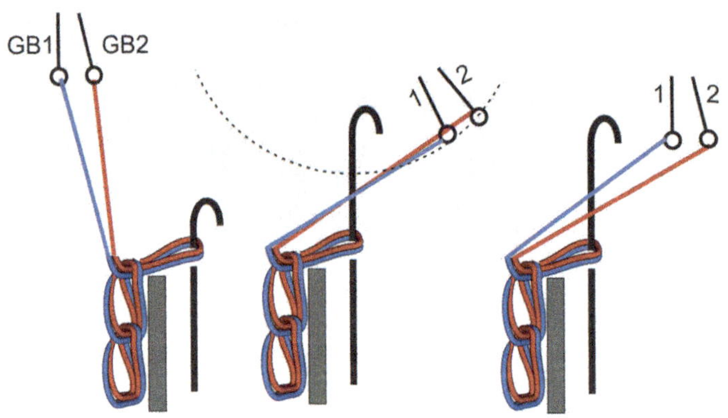

Fig. 5.13 Order of the loop head in the plating loops depends on the machine configuration and the motion of the guides. Guides with smaller radius or swinging motion

(with special feeding system). For warp knitted structures, the plating is a more or less natural process. Most warp knitted fabrics are produced with at least two guide bars working in opposite directions to improve the stability. Because of this, the loops consist of two yarns.

The arrangements of the loop elements (heads and underlaps) topologically depend on the arrangement of the guide bars, but mechanically also depend on the run-in (the yarn length per cycle), which determines the yarn tension and the pattern. According to D. Paling [26] (page 81) "It is usually the threads of the front guide bar which appear most prominently on the face of the fabric...". As D. Spencer confirms "Under normal conditions, the threads of the front guide bar tend to dominate the

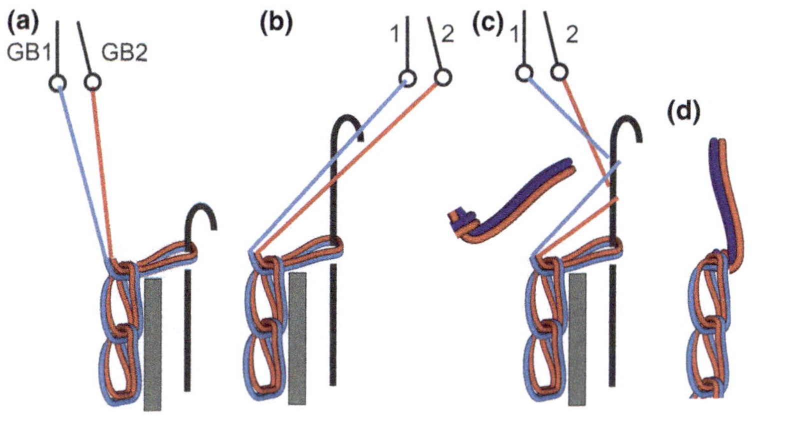

Fig. 5.14 Building of plating loops in the warp knitting—the arrangement of the guide bars determine the position of the heads and underlaps

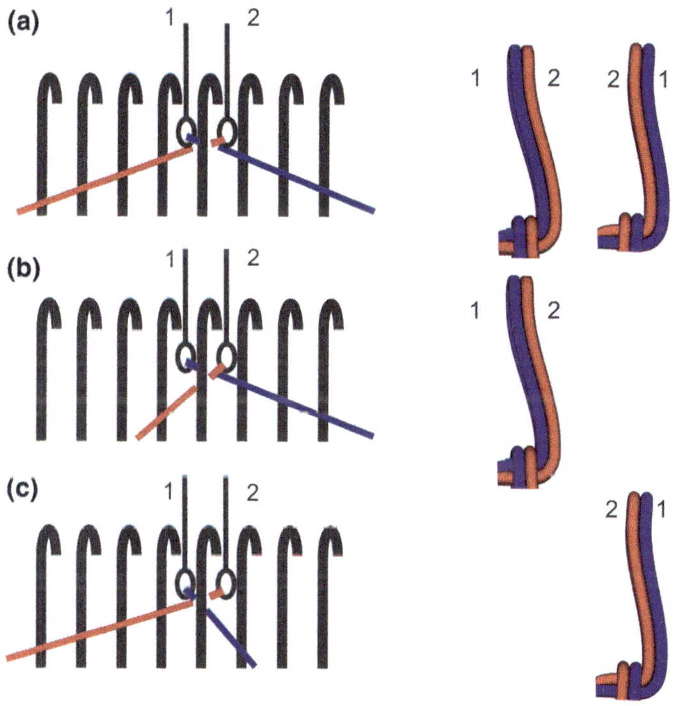

Fig. 5.15 Order of the loop head in the plating loops depends on underlap (snogging motion) before. Yarns, coming from far away have higher contact position on the needle shaft than yarns coming from the neighbour needles and this way, the order in the loop is determined

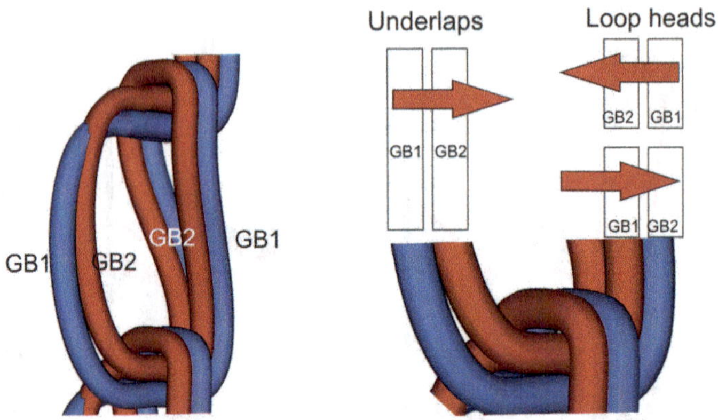

Fig. 5.16 Building of plated loops in warp knitting—the arrangement of the guide bars, but as well the pattern and the yarn tension determine the position of the heads and underlaps

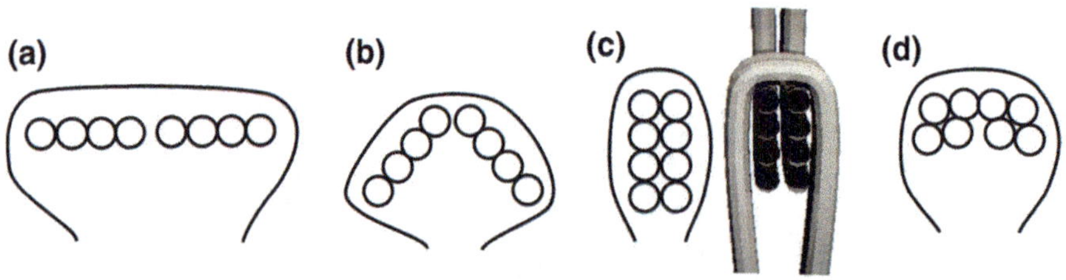

Fig. 5.17 The placement of the yarn pieces of the next loop within the head of the previous loop depends on the pattern and the yarn tension. Arrangement in horizontal line (**a**) is simple for programming but obviously not realistic for more than two yarns; configuration (**b**) can be chosen as compromise between horizontal and vertical arrangement; the vertical arrangement (**c**) leads to acceptable loop form, but to too much intersection in the yarns in the area of the underlappings (explained later). In the reality, the configuration is similar to (**d**)

face as well as the back of the fabric (Fig. 25.1)." [34] (Ch. 25, p. 313). But later Rule 3 on the same page: "A low setting of a guide, a longer run-in, an open instead of a closed lap, or a short underlap movement, will all tend to cause a warp thread to occupy a low position on the needle, either reinforcing ***or reversing*** the normal front guide bar/back guide bar plating relationship. Carefully arranged lapping movement can thus overcome the normal plating dominance of the front guide bar threads on both surfaces of the structure" [34].

These citations give the impression, that there is "normal" or "usual" plating, but this is not always the case. The modelling of the plated loops with reverse order can be done in a analogical way, as explained for the normal order. The determination of the type of the plating during the modelling can be done using some geometrical rules, based on the underlaps Fig. 5.15 and depending on the position and construction of the machine Fig. 5.13. The dependence from the run-in of the yarn, which influences the available yarn length around the loop, can not be implemented in topological

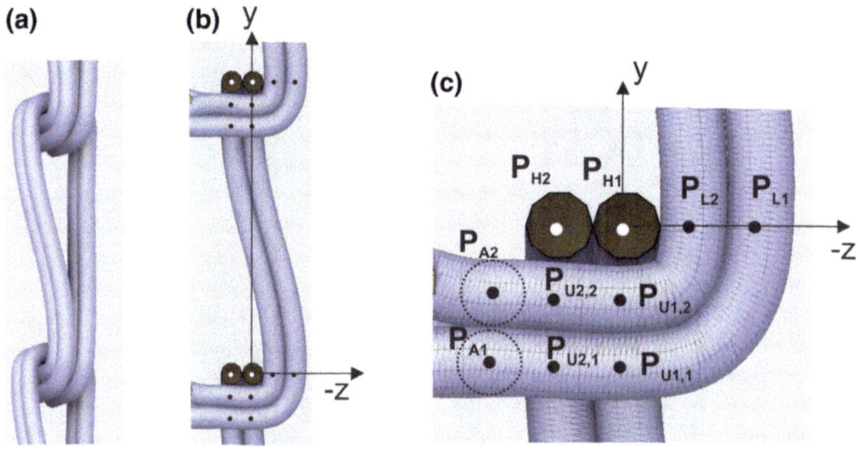

Fig. 5.18 Characteristic points for the representation of commonly produced plating loops based on two yarns, **a** isometric view, **b** view of the y-z plane (trough the loop head), **c** detailed view of the contact place between the loop head and the legs of the next loop

algorithms, as the orientation in this case depends on bending stiffness of the yarns, contact forces and friction between yarns and needles.

During the warp knitting process, the guide bars are numbered in a way, that the first bar is the closest to the needles during the overlapping (Fig. 5.14b) or the most outer one, during the underlapping (Fig. 5.14a, c). Because the yarns come from the previous loop, go over the needle and change their direction during the overlapping, the position of each loop heads over the needle can vary.

The coordinates of the yarns of both of the loops can be described as it is usually done, but with one important change—their position on the needle and in Z-direction of the 3D model is changing. The yarn from first guide bar creates the bottom loop and the next one places it in close contact with the first loop Fig. 5.18b. If the position of the head of the previous loop is already known and the coordinates of the point P_{H1} are known, then its projection over the z and y direction over the another loops can be determined as points P_{Li} and $P_{U1,i}$, where i is the number of the loop, Fig. 5.18c, and in this way the head with the plating two loops is already described:

$$\mathbf{P_{L2}} = \mathbf{P_{H1}} - \mathbf{e_z} \cdot (r_1 + r_2) \tag{5.12}$$

Here and during the next consideration is made an assumption, that in one loop head the yarns from the next loops are arranged vertically, as demonstrated in Fig. 5.17c). The real arrangement of the yarns in the loops depends on the lapping, yarn tension and yarn properties and for the case of plating of four yarns can look like Fig. 5.17d). The definition of some parametric equations for such one configuration require large algorithm with several cases. The horizontal orientation (Fig. 5.17a) is simple but not realistically, because the yarn stack will collapse.

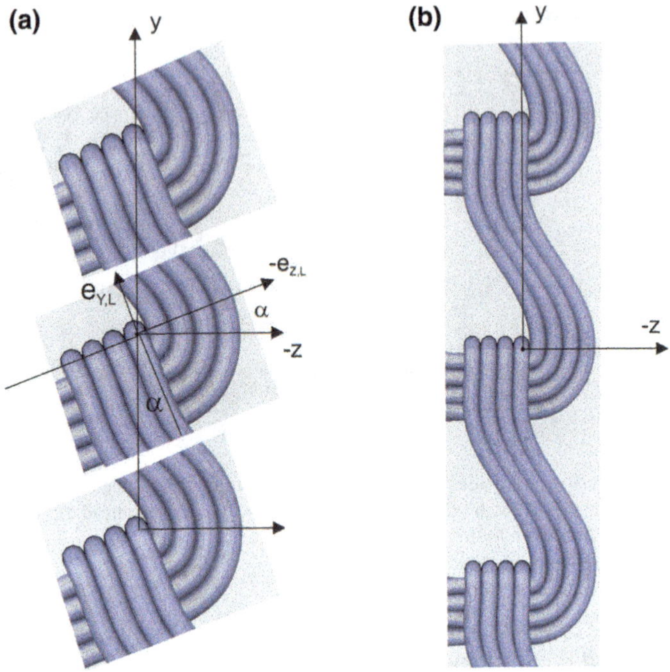

Fig. 5.19 Plating loops with more yarns. **a** Geometrically correct model, **b** simplified, topologically correct representation

For plated loops with more yarns the complete stack of the loop heads becomes larger (Fig. 5.19a). It can be considered as a set of parallel loops, which are all oriented in angle α to the vertical $(0y)$ axis. The angle α depends on the number and diameters of the the yarns in that loop and the loop height. Four yarns per needle are not used very often and therefore, if the model has only be correct topologically, an alternative representation with vertical loop heads can be used (Fig. 5.19b). It is more simplified in the realization because do not need the rotation of the stack of yarn pieces.

In both cases, the characteristic points of each yarn can be defined based on the coordinates of the loop head P_{H1}, where in the case of rotated head, a local coordinate system with unit vectors $\mathbf{e_x}$, $\mathbf{e_{Y,L}}$, $\mathbf{e_{Z,L}}$ has to be used (Fig. 5.20) additionally. The x coordinates of all points can be determined analogically to these of the single yarn loops.

In reality, yarns with different finesse (resp. diameter) are often used in the same structure. Because of this, it is more efficient to not choose the loop head point $\mathbf{P_{H1}}$ as the origin of the calculations of the loop head coordinates, but to choose the point in the loop below it, that is positioned within the loop head—$\mathbf{P_{U1,M}}$. Starting from this point, the distance of the projection of the next yarn center can be found along the axis $\mathbf{e_{Z,L,II}}$ and then the yarn center can be found adding a vector in direction

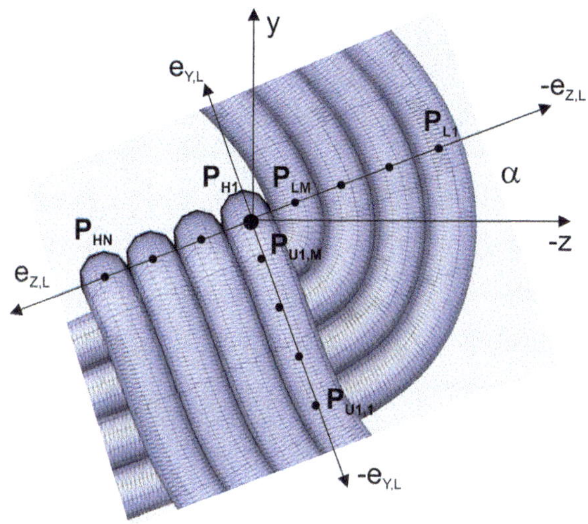

Fig. 5.20 Plating loops with three loops

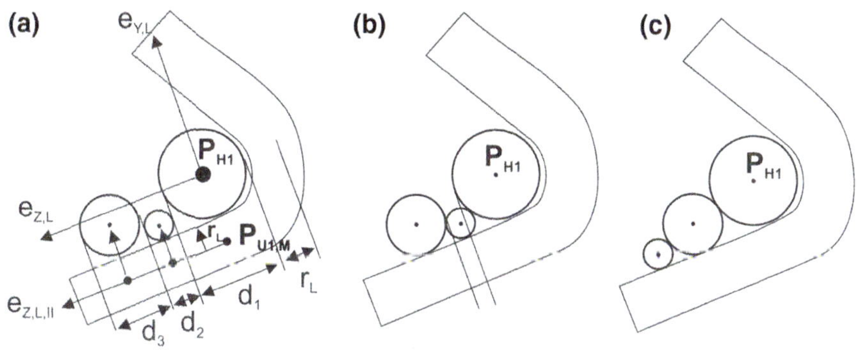

Fig. 5.21 Simplified (**a**), geometrically correct (**b**) and some times real configuration (**c**) of three loop heads of different yarn diameter in one plated loop

$\mathbf{e}_{Y,L}$ at distance to one $r_L + r_i$, where r_L is the radius of the yarn of the new loop and r_i is the radius of each yarn, which loops are held by the current one. Here, in practice, the projections of the yarn diameters do not provide an accurate equation for determination of the contact between the yarns, because the thinner yarns can partially be placed **under** the thicker yarns Fig. 5.21b or even worst—based on the different bending stiffness or yarn tension they can partially move close to the next yarn, Fig. 5.21c.

If the correct coordinates of the loops have to be created, the geometry, presented in Fig. 5.22 can be applied. There are different ways for determining the position of the point $\mathbf{P_{H2}}$, based on the $\mathbf{P_{H1}}$. Mathematically, the point $\mathbf{P_{H2}}$ can be found as a solution of a non linear system of equations, based on the distance between it. Its

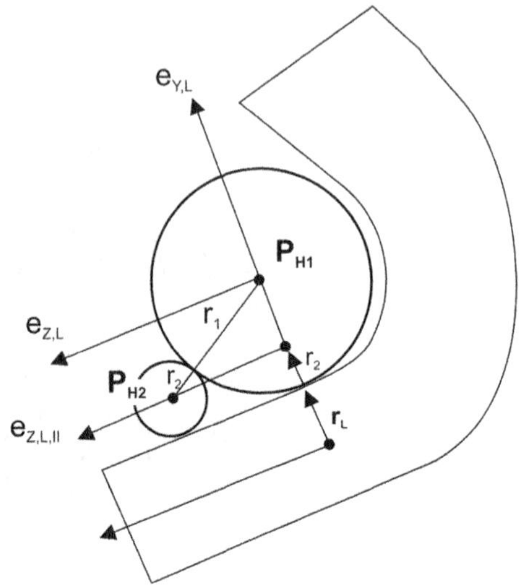

Fig. 5.22 Geometric model for correct calculation of the position of the yarns with different thickness in the loop

place can be found as a crossing point of the circle with radius $r_1 + r_2$ around $\mathbf{P_{H1}}$ and the vector $\mathbf{e_{Z,L,II}}$. Actually, the nonlinear systems require iterative procedures and do not always lead to the correct solution. In this case, the solution can be found based on the geometry more easily: the origin of the vector $\mathbf{e_{Z,L,II}}$ can be calculated as

$$\mathbf{P_{H1,st}} = \mathbf{P_{H1}} - \mathbf{e_{Y,L}} \cdot (r_1 - r_2) \tag{5.13}$$

The distance between $\mathbf{P_{H1,st}}$ and $\mathbf{P_{H2}}$ is

$$\|\mathbf{P_{H1,st}} - \mathbf{P_{H2}}\| = r_1 + r_2 - \sqrt{r_1 - r_2} \tag{5.14}$$

And thus, the point $\mathbf{P_{H2}}$ is determined. The same procedure has to be repeated for the next pair of cross sections of yarns on the loop head. As mentioned, this accuracy is barely efficient in the modelling process, because the knitted structures show significantly larger deviations in their form, based on the friction, yarn length and yarn tension differences, so any improvement in the accuracy of the placement of the heads of the loops in the plating loops will not lead to remarkable improvements in the general appearance for large sets of variations.

The previous considerations was (considerations were done OR consideration was done) done under the assumption, that it is known how the loops are placed, which depends on the arrangement of the guide bars. There are often situations in samples, where some guides are empty (partial threading) and do not place the yarn over each needle. In such a case, some loops contain more yarns, other less. Figure 5.23 demonstrates such a situation—the loops in the first and third columns are built from two yarns, and the middle one from four yarns. In order to keep the topologically correct arrangement of the loops, they can be placed one after another, depending

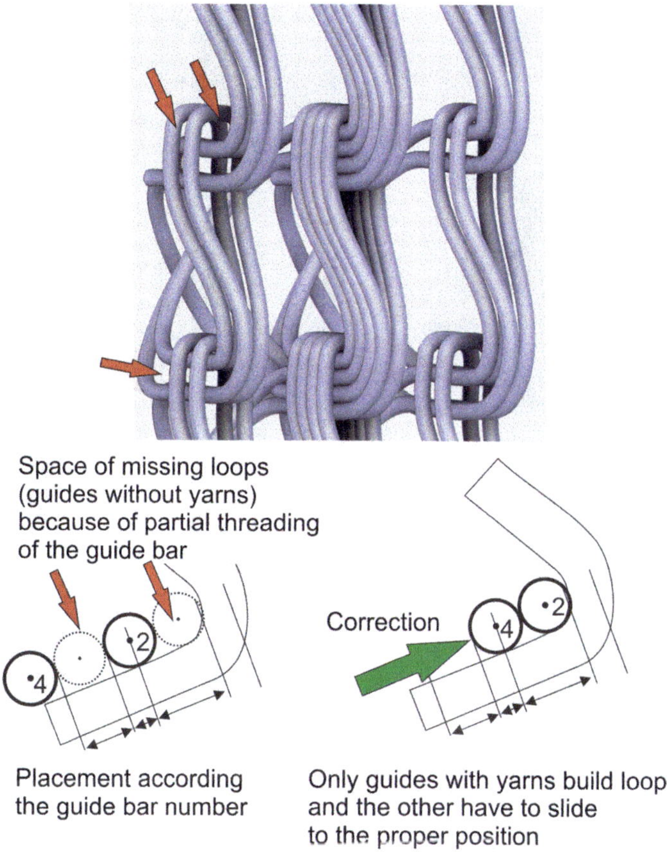

Fig. 5.23 Spaces between loops at partial threading and possible way for their correction

on the number of the guide bars. But if in this case, the guide bars 2 and 4 have yarns, the positions of the first and third will remain unused. For all these situations, the positions of the loops have to be adjusted in $-e_Z$ direction. This is one of the corrections, which can be done during the second, "corrector"-step of the simulation. Initially—during the first step—it is not known, which yarns with which diameter are placed on which position.

5.7.1 Underlaps in Plated Loops

Another problem, which is very common concerning the topology of the plated loops, is the orientation of the underlaps. This systematical problem is occurring for warp knitted structures (Fig. 5.24).

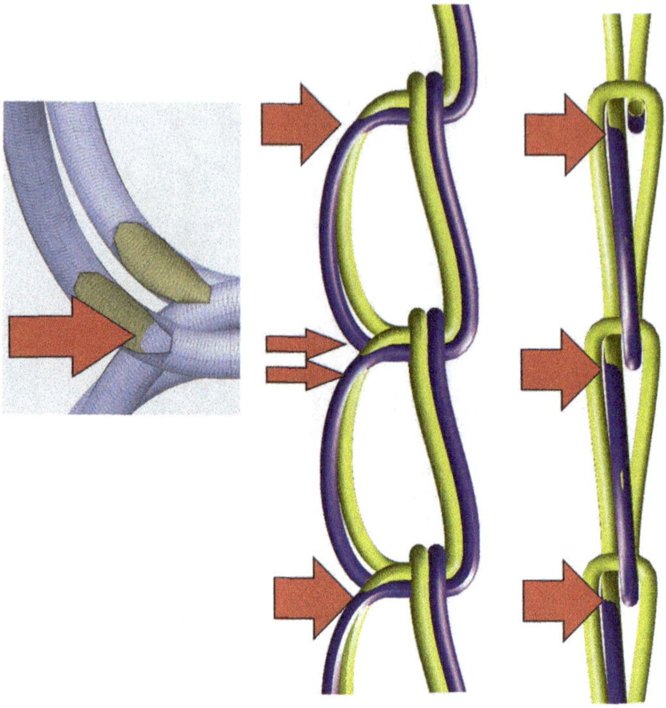

Fig. 5.24 Intersection of the underlap for 1-0/1-0// pattern. **a** Magnified view with section trough the yarns, in order to make the interpenetration visible, **b** side view and **c** front view of the modeled plated loops, where the underlaps are intersecting

The underlaps connect two loops and in weft knitted structures and are just a horizontal piece of yarn. For warp knitted structures, the underlaps are what the knitter sees from one side. When producing a fabric, they connect the loop from one wale to the next one, but normally each guide bar moves to a different direction. At one type of definition of the coordinates for one type of loops (Fig. 5.25 left) can lead to acceptable geometry, but small change in the pattern (replacing the open loop with closed) can lead to massive intersections between the underlaps (Fig. 5.25 right hand side).

A common and simple situation is the lock-stitch loop of two guide bars, doing the same pattern. The underlaps change their position Fig. 5.24.

The analysis of different configurations for *one* guide bars shows, that for structures, based only on loops, there are 18 different possible configurations of the two underlaps around one loop, depending on the type of the previous loop, current one and the next one (Fig. 5.26). Each configuration is depending on three cycles, which means, on three numbers.

Let each step of the motion of the guide be coded as variable T, depending on the difference between the swinging positions of the guide bar during the swing-in and swing-out motion parts. There are three numbers, which determine the orientation of the underlaps around one loop:

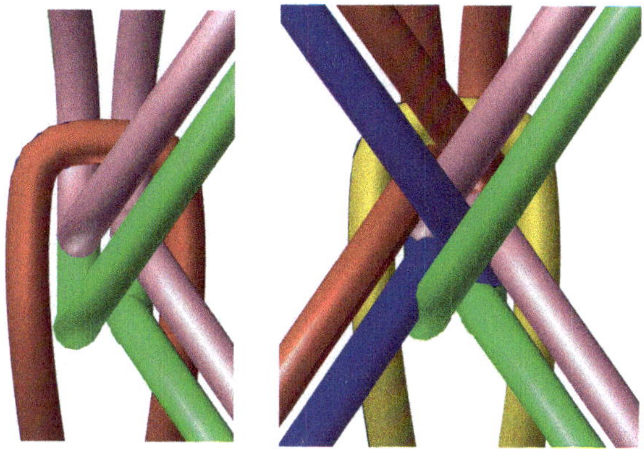

Fig. 5.25 Underlapping for plated loops in collision free configuration for one pattern and intersecting at other pattern

Incoming—underlap

$$T_{in} = in - out_{last\ loop} \tag{5.15}$$

the loop

$$T = out - in \tag{5.16}$$

and the outgoing underlap

$$T_{out} = in_{next\ loop} - out \tag{5.17}$$

If only "loops" are allowed, T can only have two values: +1 and −1; and T_{in} and T_{out} can have three values each −1, 0, or +1 (−1 for moving guide bar in negative direction, 0 for noch changing the direction, +1 for moving in positive direction). The total number of combinations is $3 \cdot 2 \cdot 3 = 18$. As Fig. 5.26 demonstrates, for a plated loop made from two yarns, for each of these 18 possible orientations of the underlaps 18 configuration for the underlaps of the another plated loop are possible. This means, that in case of two guide bars, there are $18 \cdot 18 = 324$ configurations for the orientations of the underlaps. The modern machines have often 4 some more than 10 guide bars. Several of them are not fully threaded, but still the number of the combinations of the orientations of the underlaps are too large for considering all of them in one algorithm. Any trial of defining a geometrically correct model, which avoids interlacement of these yarns, has to consider all these configurations, or even more, if more guides have yarns which are overlapping one and the same needle. The principle for the definitions of such configurations is similar to those, of the orientation of the loops heads, and will not be discussed further.

More important for the modelling is, that each loop or each structural element can not be determined in the space, without the information about the orientation of the previous and the *next* one (Fig. 5.27).

Looking at this triple dependency, it becomes clear, that each algorithm for the creation of topologically correct representations of knitted loops *has to* consists of

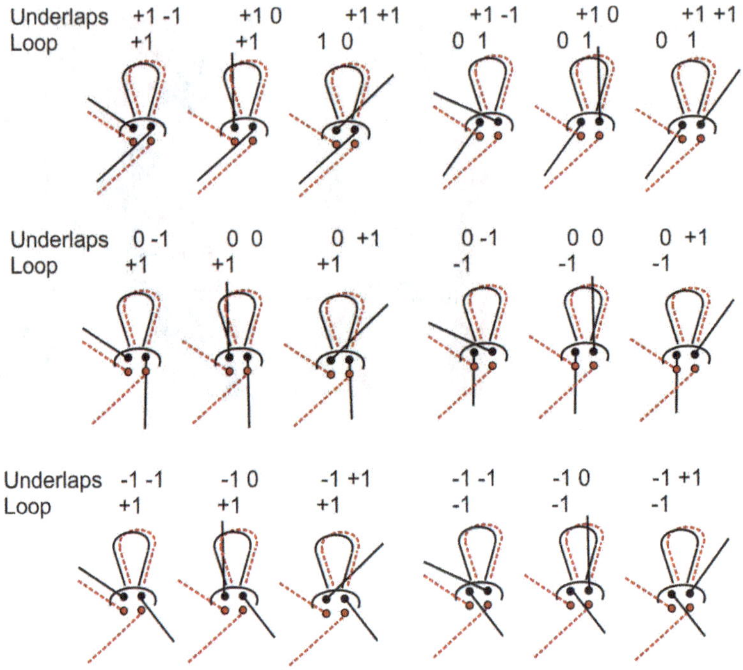

Fig. 5.26 All possible combinations of the orientation of the underlaps of the yarn at one guide bar (solid yarn), if the second guide bar makes always the same type of loop (red dashed line)

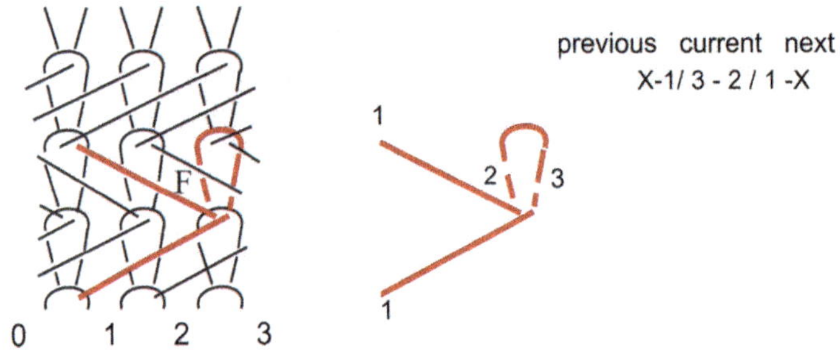

Fig. 5.27 The configuration of each loop depends on the code of itself, the type and position of the previous one and the type and the place of the next one

two virtual knitting steps, which are identical to the *predictor-corrector* methods of solving differential equations (Fig. 5.28):

- first step, the *predictor*, initializes all structural elements, sets their types, positions, number of elements in each knitting place
- second step, the *corrector* has to adjust the positions of the single elements, based on the data, already available.

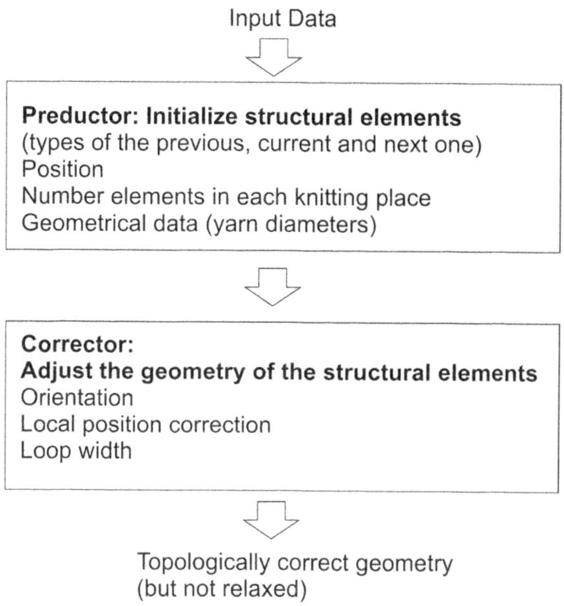

Fig. 5.28 Two step, predictor-corrector algorithm is required in order for the position and orientation of the structural elements of knitted structures to be calculated

5.7.2 *Transferred Loops*

The loop transfer is typical for the weft knitting machines and is seldomly used in modern warp knitting (an exception is given in [26]). Transferred loops (Fig. 5.29(T)) are produced on one needle and after that are moved to some other needle. The new loop on the needle, where the moved loop has been made before transferring, has a different form (Fig. 5.29(N)). It is similar to single tuck, without loop. This has to be considered during the modelling of the geometry. There are advanced structures with higher complexity, e.g. braids, which require more sophisticated treatment.

5.8 Plated Loops in Double Needle Bed Structures

As their name says, the double needle bed machines have two needle beds, which build the loops. In some cases—always in the warp knitting and for some structures in the weft knitting—the two beds are working one after another—the needles of the front bed build loops and afterwards the needles of the rear bed. In general case of the weft knitting (except when tubular knits are produced), the needles of both needle beds build the loops simultaneously.

Generally speaking, the loops and all other structural elements in the rear bed should have a mirrored geometry to these in the front bed, as these beds are placed symmetrically. This rule is true for the weft knitted structures, but is not true for the

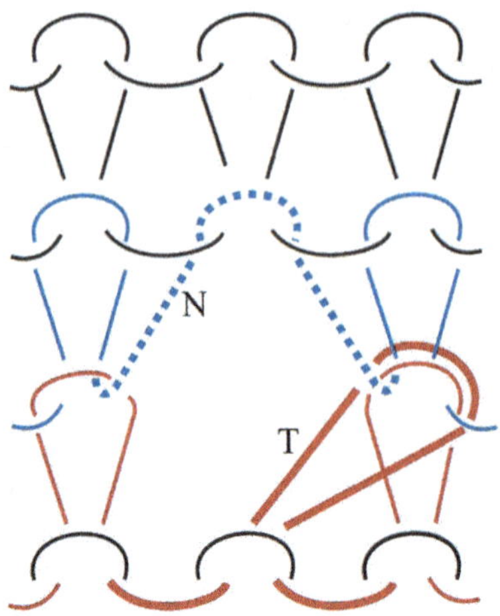

Fig. 5.29 Transferred loop T and the new build tuck N, which will be used as a basis for later loop. Reprinted from X. Chen (Ed.) "Modelling and predicting textile behaviour", Y. Kyosev, W. Renkens, "Modelling and visualization of knitted fabrics", pp. 225–262, Copyright (2010) with permission from Elsevier [15]

warp knitted structures. On the double needle bed of warp knitting machines, the guide bars keep their arrangement during the entire process and because of this, the front guide bar works as front bar for the front needle bed, but it becomes the rear (or the last) bar for the rear needle bed (Fig. 5.30)

Keeping in mind, that the arrangement of the underlaps always depends on the number (and position) of the guide bar, the rules for their positioning in the double needle bed structures can derived as extension of the rules for plating loops for single needle bed ones. In the most of the cases, the arrangement at the upper line of the figure (Fig. 5.30) is valid, but for different yarn tensions and some other patterns, the bottom line can become valid. For some pattern all four combinations for different elements can be valid, too.

5.9 Weft Insertion

The weft insertion is a normal element in warp knitted structures. It is used to stabilize the structure in horizontal directions (Fig. 5.31) or for building net shaped structures (not presented here). It is a very important structural element from a mechanical point of view, as it changes the flexibility resp. elongation of the structure drastically. It is used in light nets as well as in structures for textile reinforced composites, where heavy tows (usually carbon or glass) are inserted as weft yarns. In warp knitted structures like these, the only task the loops have, is to keep these tows together. There are two different subtypes of the weft insertions—"partial" weft, not going trough the complete width and "full" or magazine weft, which is placed from a separate device completely trough the entire width of the warp knitted product. Topologically are these two subtypes identical, but the full weft yarns are taking different position

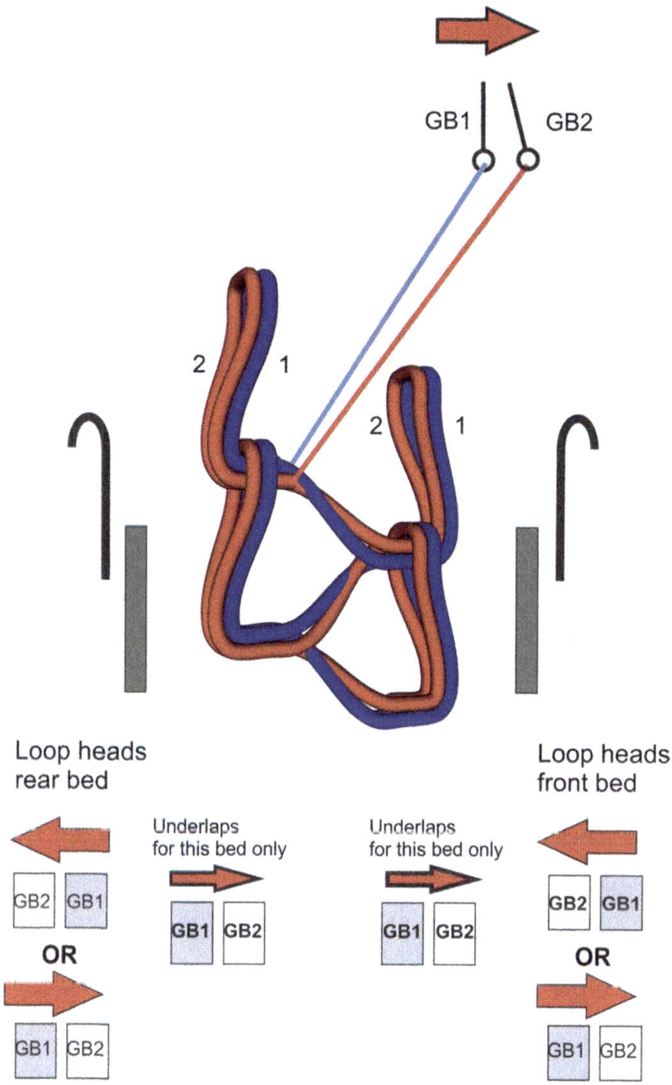

Fig. 5.30 The orientation of the loops in the rear bed remains the same, as in the front one, because the guide bars do not change their positions

in the loop, because of the differences in the placement geometry and forces. These differences will be not discussed more here.

The description of the points of the weft insertion can be described using a plane projection of the yarn, as it normally is placed in one plane. This plane is placed between the loop legs and the underlaps. An accurate description of the weft insertion will consist of arc and connecting lines. Their coordinates can be derived on the base

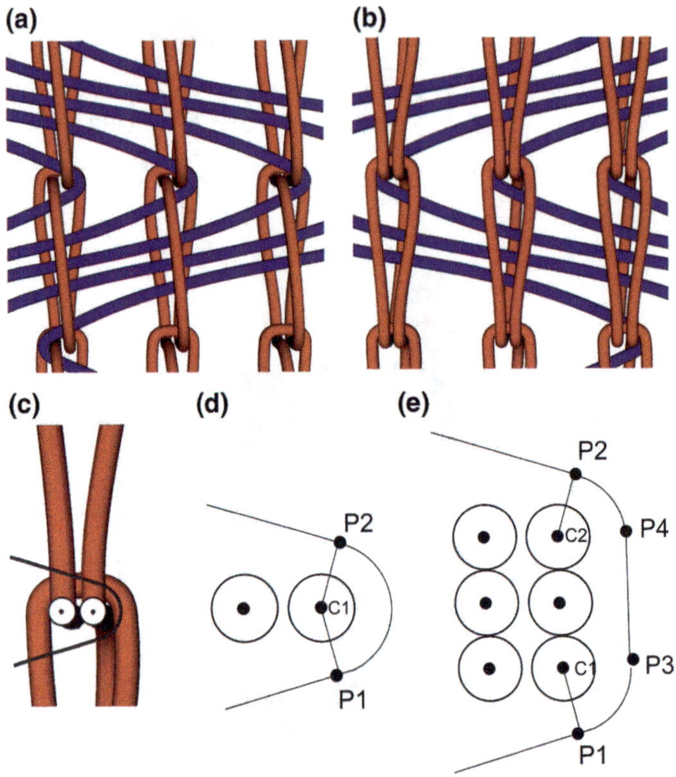

Fig. 5.31 Weft insertion and principle of its geometrical description **a** and **b** view from technical front and technical back of the structure, **c** principle representation of the structural element, **d** characteristic points, **e** characteristic points in case of weft yarn running around plated loops with multiple yarns

of the crossing point of the loop yarn axis with the plane of the weft insertion C_1 and the information about the yarn radii. In case of plating loops, the arc should be split in two and the most upper and most bottom points C_1 and C_2 then should be considered. For the weft insertions, again, this method only presents topologically correct representations, defined using a limiting position of the yarns. In reality, the exact position of the weft yarn inserted, depends on the run-in (its length) or its tension respectively (the run-in define the available yarn length and in this way the yarn tension). As depicted in Fig. 5.32 the weft yarn can as well be placed not exactly contacting the loop legs, because after the lapping movement it can relax and takes some middle position between the loops, if the length for this is available. The information about the yarn length or its tension is normally not available during the modelling, so the topological representation gives the correct interlacement with the other elements only.

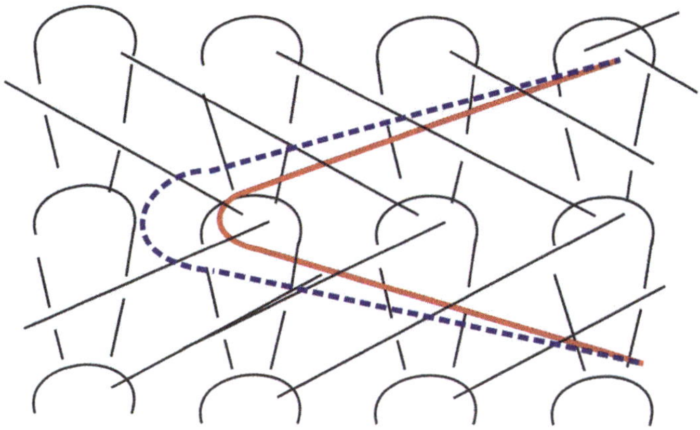

Fig. 5.32 Position of the weft yarn depending on the yarn tension

5.10 Weft and Warp Yarns Reinforced Knitted Structure (MLG)

For technical applications the UD (Uni-directional) structures are very common used. Uni-directional fabric is a layer of parallel strong heavy tows or tick rovigns, which are kept together in with some additional thin yarn system. Most used kind of such fabrics belongs to the group of warp knitted structures and their modelling is very well described in the work of Robitaille et al. [30]. Software implementation of such structure is available independent on this work as part of Wisetex under the group of Stitched UD fabrics [18]. Less developed (at year 2005, from which the work of the author is) were the models for the weft knitted such structures. The development of a modified machine, method for production and set of products of weft knitted structures were developed by Cebulla [1]. The weft knitting allows production of structures with predefined shape and with more intelligence and computations as well 3D form, which is not possible with the warp knitting. In the current case the thicker warp and weft yarns are kept together into structures from a thin knitting yarns (Fig. 5.33). The model for the points considers, that such fabrics are produced from thick warp and weft multifilament yarns. These yarns have flat, elliptic or lenticular cross section with width and height (Fig. 5.34) [12]. As the warp and weft yarns are significantly thicker, they remain straight and their position is defined by any two points of their axis. The axis of the loop yarns is in this case not following the common model for knitted structures, because the loop yarn is placed around the weft and warp yarns. The main points of it are visualised in Fig. 5.35.

For this section only the following indexes are used—subscript M—indicate the loop yarn (M = Masche in German); K = warp yarn (K = Kette), S = Weft yarns (S = Schuss) 1 is the smaller half thickness, “radis” and 2 the larger half thickness, or half width.

$$K_1M_1 = r_{K1} + r_{M1} \tag{5.18}$$

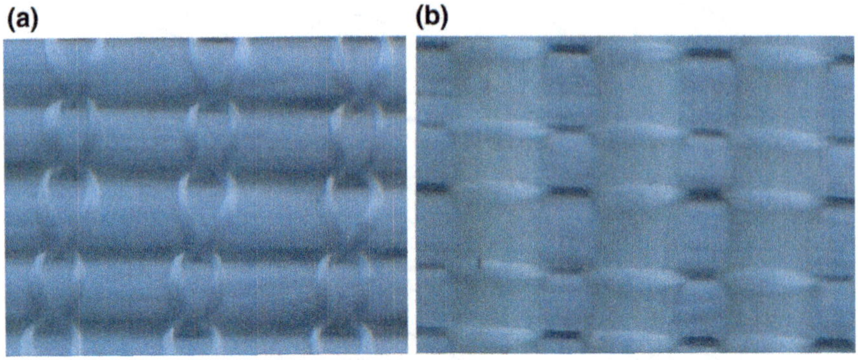

Fig. 5.33 Weft knitted structure with weft and warp yarns **a** loop side, **b** sinker loop side [12]

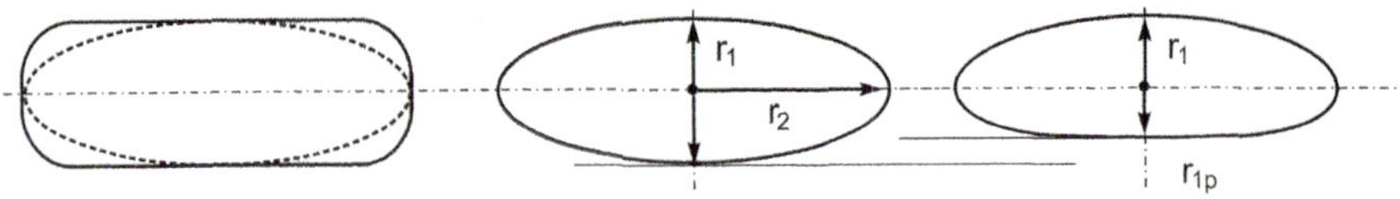

Fig. 5.34 Assumption about the yarn cross section

For the coordinate system is selected OX axis parallel to the weft yarns, OY axis, parallel to the warp yarns, and Z direction is determined by right hand oriented coordinate system.

The following points can be determinated from geometrical relations:

$$K_1 = \left[0\ 0\ 0\right] \tag{5.19}$$

$$S_1 = [0\ 0\ r_{K1p} + r_{S1p}] \tag{5.20}$$

$$M_1 = [0\ y_{M1}\ r_{K1} + r_{M1}] \tag{5.21}$$

Here $y_{M1} = y_{M3} - \Delta h$ and at the point $M2$ the loop lost the contact to the warp yarns.

$$M_2 = [r_{K2} + r_M\ y_3 - \left(r_{K1p} + r_{S1p}\right) \cdot \tan\alpha\ 0] \tag{5.22}$$

$$\tan\alpha = \frac{\Delta h}{r_{K1} + r_{K1p} + r_{S1p} + r_M} \tag{5.23}$$

Δh is the distance between the y-projections of the points M_3 and M_1, or technologically it is the approximated distance between the middle of the opening between the weft yarns and the axis of the loop over the warp yarn

At the point M_3 the loop yarn is already over the weft yarn (Fig. 5.36)

$$M_3 = [r_{K2} + r_{M1}\ r_{S2} + r_{M2}\ r_{K1p} + r_{S1p}] \tag{5.24}$$

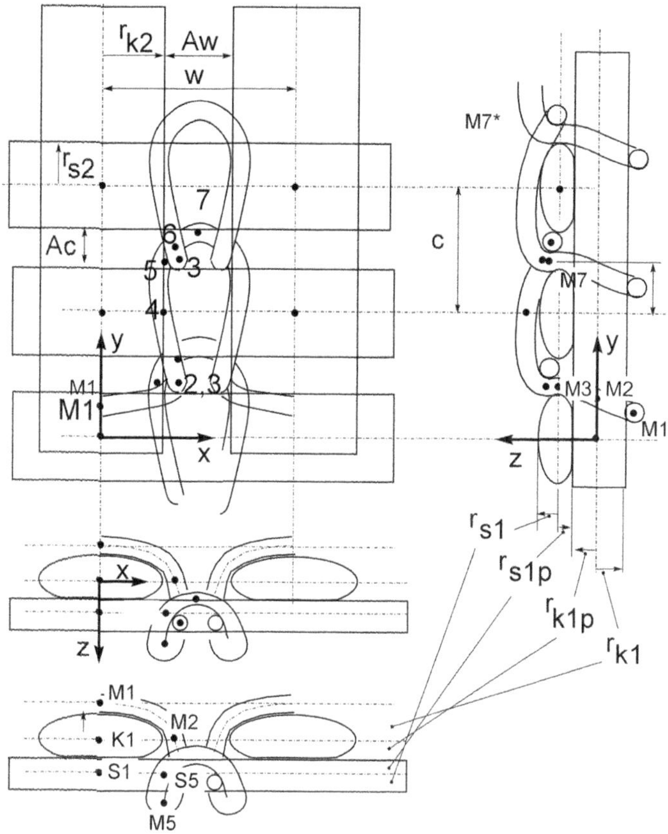

Fig. 5.35 Geometrical model of the structure

Fig. 5.36 Determination of the point Me

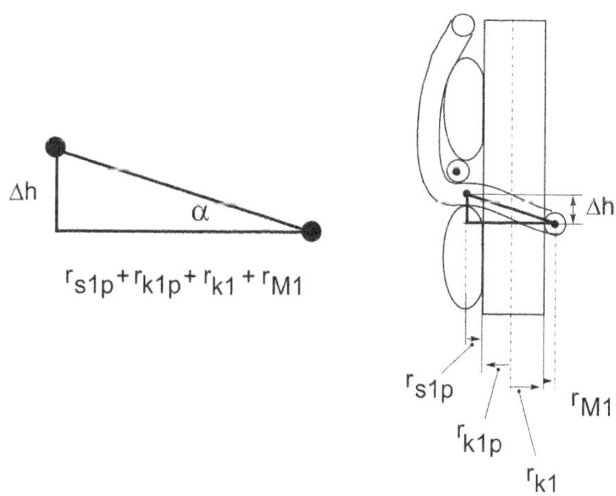

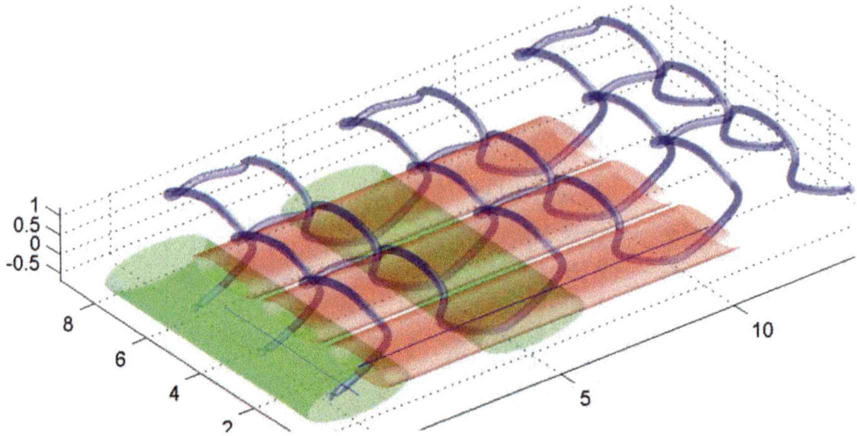

Fig. 5.37 Simulated fabric, visualised within Matlab with transparent yarns [12]

$$M_4 = [\, r_{K2} + x_{4K} \;\; c \;\; r_{K1p} + r_{S1p} + r_{S1} + r_{M1}] \tag{5.25}$$

For the point 4 $x_{4K} = 0$ at the beginning. The exact value is calculated later.

$$M_5 = [\, x_{M3} - 2 \cdot r_{M1} \;\; y_{M3} + c \;\; z_{M3}] \tag{5.26}$$

$$M_6 = [\, x_{M3} \;\; y_{M3} + c + r_{M1} \;\; z_{M3} - 2 \cdot r_{M1}] \tag{5.27}$$

Point M7 is the middle of the loop arc:

$$M_7 = [\, w/2 \;\; c + r_{K2} + r_{M1} + y_{M7K} \;\; z_{M6} + z_{M7K}] \tag{5.28}$$

For the next points the loop is symmetrical, so the coordinates are calculated as follow:

$$x_{i,sym} = w - x_i \tag{5.29}$$

The resulting geometry is visualised in the Fig. 5.37. The parametrical model allow as well calculation of the length and geometry changes up to the jamming state in similar way as it is done for the braided structures. The jammed geometry with consideration of the thickness of the loop-yarns is visualised in Fig. 5.38a and if the thickness of the loop-yarn is neglected on Fig. 5.38b.

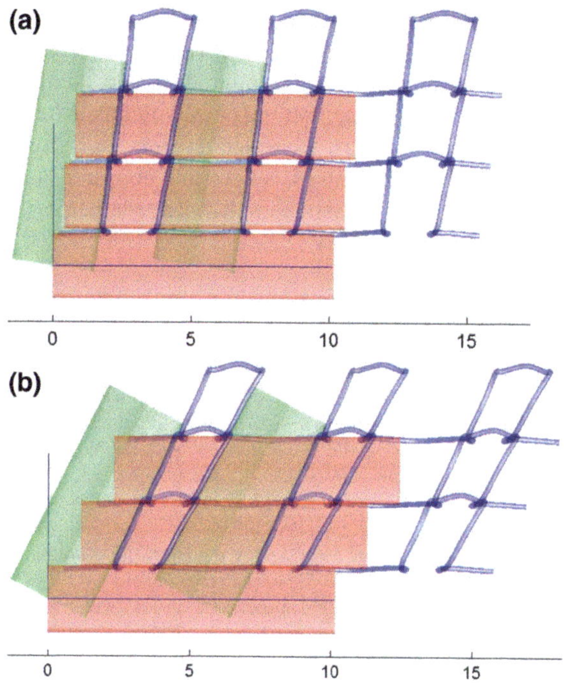

Fig. 5.38 Jammed states of the fabric during shear loading **a** with consideration of the loop building yarn thickness, **b** ignoring it

5.11 Other or Modified Structural Elements

The inlays are yarn (pieces) received from a warp beam, which are normally located between the floats, the loops and the weft insertions. They usually have a higher thickness than the yarns, building loops, and for this reason they remain almost in straight form. The yarns from full weft systems do also have a almost straight form. These can be inserted horizontally or under some angle, prepared on a magazine. The yarns of the full wefts are often split into two parts, if the needles cross them during the knitting processes.

In some special types of structures, modified elements can be found. For instance, pile samples (plush) contain loops with higher lengths. Sometimes they are cut (velvet) and sometimes they build a relief surface.

5.12 Conclusions

The definition of the single structural elements of the knitted structures using relations for the position of their specific points, allow quick description of the 3D yarn geometry of these structures. Although the exact geometry of the loops differs significantly from the initially generated form, these methods allow quick and clear topological description of the elements and in form of a complete structure. Therefore, the generated geometry is already useful for the textile engineers, as it allows rapid 3D visualization of complex lists with numerical data for the motion of the guides in case of warp knitting or of commands for the control of the needles in the weft knitting. In cases, for which this topologically correct models are not accurate enough, these can be used as a starting point for more accurate mechanical models, using the methods described in the next chapters.

References

1. Cebulla, H.: Formgerechte zwei- und dreidimensionale Mehrlagengestricke mit biaxialer Verstärkung: Entwicklung von Maschine, Technologie und Produkten, Dresdner Forschungen Maschinenwesen, vol. 20. TUDpress, Verl. der Wiss, Dresden (2005)
2. Chen, X., Hearle, J.W.S.: Developments in design, manufacture and use of 3D woven fabrics. In: Advani, S.G., Gillespie, J.W. (eds.) Recent Advances in Textile Composites. DEStech Publications, Lancaster, PA (2008)
3. Choi, K.F., Lo Y, T.: An energy model of plain knitted fabric. Text. Res. J. **73**, 739 (2003). https://doi.org/10.1177/004051750307300813
4. Doyle, P.J.: Fundametnal apspects of the design of knitted fabrics. J. Text. Inst. **44**, P561–P578 (1953)
5. Grishanov, S., Lomov, S.E.A.: The simulation of the geometry of two-component yarns. Part I: the mechanics of strange compression: simulating yarn cross-section shape. J. Text. Inst. **88**((Part 1)(2)) (1997)
6. Harwood, R., Grishanov, S., Lomov, S., Cassidy, T.: Modelling of the two-component yarns: part 1—the compressibility of yarns. J. Ind. Text. **88**(Part 1.4), 373–384 (1997)
7. House, D., Breen, D. (eds.): Cloth Modelling and Animation. A K Peters, Natick, Massachusetts (2000)
8. Kaldor, J., James, D.L., Marschner, S.: Simulating knitted cloth at the yarn level. In: Proceedings of SIGGRAPH 2008. Held in Los Angeles, California, Aug 2008
9. Kurbak, A.: More about the rib knitted fabric dimensions. Ege University, Textile Engineering Department, Izmir (1995)
10. Kurbak, A., Ekmen, O.: Basic studies for modeling complex weft knitted fabric structures part I: a geometrical model for widthwise curlings of plain knitted fabrics. Text. Res. J. **78**(3), 198–208 (2008). https://doi.org/10.1177/0040517507082352
11. Kurbak, A., Soydan, A.S.: Basic studies for modeling complex weft knitted fabric structures part III: a geometrical model for 1 x 1 purl fabrics. Text. Res. J. **78**(5), 377–381 (2008). https://doi.org/10.1177/0040517507082465
12. Kyosev, Y., Haasemann, G., Diestel, O., Orawattanasrikul, S.: Geometrical modelling tool for biaxial reinforced weft knitted structures. In: Proceedings of Aachener Textile Conference, Aachen (2005)
13. Kyosev, Y.K.: Texmind warp knitting editor 3D (2017). www.texmind.com
14. Kyosev, Y.K., Angelova Y., Kovar, R.: 3D modelling of plain weft knitted structures from compressible yarn. Res. J. Text. Appar. **9**(1), 88–97 (2005) (Hong Kong)
15. Kyosev, Y.K., Renkens, W.: Modelling and visualization of knitted fabrics. In: Chen, X. (ed.) Modelling and Predicting Textile Behaviour, pp. 225–262. Woodhead Publishing and In association with the Textile Institute and CRC Press, Cambridge, Boca Raton, FL (2010)
16. Kyosev, Y.K., Renkens, W., Reiners, P.: Warp knitted structures—geometrical and mechanical modelling. In: 46th International Congres IFKT, Sinaia, Rumania (2012)
17. Lomov, S., Bernal, E., Koissin, V., Peeters, T.: Integrated textile preprocessor wisetex, version 2.5: computational models, methods and algorithms

18. Lomov, S., et al.: Wisetex (2011). http://www.mtm.kuleuven.be/Onderzoek/Composites/Research/meso-macro/textile_composites_map/textile_modelling/textile_modelling_fe
19. Lomov, S., Gusakov, A., Huysmans, G., Prodromou, A., Verpoest, I.: Textile geometry preprocessor for meso-mechanical models of woven composites. Compos. Sci. Technol. **60**(11), 2083–2095 (2000)
20. Meißner, M., Eberhardt, B.: The art of knitted fabrics, realistic & physically based modelling of knitted patterns. In: EUROGRAPHICS '98, vol. 17, no. 3 (1998)
21. Meissner, M., Eberhardt, B., Strasser, W.: A Volumetric Appearance Model: Cloth Modeling and Animation, pp. 269–285. A K Peters, Natick, Massachusetts (2000)
22. Moesen, M.: Modelleren van de vlakke vervorming van gladde inslagbreisels. Master's thesis. http://liegebeest.studentenweb.org/weftknitEN.html
23. Moesen, M., Lomov, S., Verpoest, I.: Modelling of the geometry of weft-knit fabrics. In: Techtextil (Hg.) 2003—TechTextil Symposium, 7–10 Apr 2003
24. Moesen, M., Lomov, S.V.: Weftknit (2011). http://liegebeest.studentenweb.org/weftknitEN.html
25. Munden, D.: The geometry and dimensional properties of plain-knit fabric. J. Text. Inst. **50**, T448–471 (1959)
26. Paling, D.: Warp Knitting Technology. Notthingham (1965)
27. Postle, R., Carnaby, G., de Jong, S.: The Mechanics of Wool Structures. Ellis Horwood Limited (1988)
28. Renkens, W., Kyosev, Y.: Geometrical modelling of warp knitted fabrics. In: Finite Element Modelling of Textiles and Textile Composites, CD-ROM Proceedings, St-Petersburg, 26–28 Sept 2007
29. Renkens, W., Kyosev, Y.: Geometry modelling of warp knitted fabrics with 3D form. Text. Res. J. **81**(4), 437–443 (2011). https://doi.org/10.1177/0040517510385171
30. Robitaille, F., Clayton, B.R., Long, A.C., Souter, B.J., Rudd, C.D.: Geometric modelling of industrial preforms: warp-knitted and multiple layer textiles. J. Mater. Des. Appl. Proc. Inst. Mech. Eng. (Part L) **214**, 71–90 (2000)
31. Sherburn, M.: Geometric and mechanical modelling of textiles. Ph.D., University of Nottingham, Notthingham, July 2007. http://etheses.nottingham.ac.uk/303/1/thesis-final.pdf
32. Shima Seiki: Sds-one apex (2011)
33. Sobotka, L.: Bindepunktc als mittel zur darstellung von einfaden-gestrickbindungen. In: XLII Congress of the International Federation of Knitting Technologies (IFKT), Poland (2004)
34. Spencer, D.: Knitting technology. dd **1** (2001)
35. TexEng Software Ltd: Knitgeomodeller (2011). http://www.texeng.co.uk/knitgeomodeller.html
36. Verpoest, I., Lomov, S.V.: Virtual textile composites software wisetex: integration with micro-mechanical, permeability and structural analysis. Compos. Sci. Technol. **65**(15–16), 2563–2574 (2005)
37. Weber, K.P., Weber, M.O.: Wirkerei und Strickerei. Deutscher Fachverlag (2008)
38. Wilkens, C.: Warp knit fabric construction. U. Wilkens Verlag, Heusenstamm, Germany (1995)
39. Wu, W., Hamada, H., Maekawa, Z.: Computer simulation of the deformation of weft knitted fabtrics for composite materials. J. Text. Inst. **85**(2), 198–214 (1994)

Chapter 6
Truss Framework Model for Warp Knitted Structures

6.1 Warp Knitted Structures as Truss Framework

The warp knitted structures are based mainly on yarn loops (with heads and legs), connected by underlapps. The loop heads are adjusting their place in the space according to the machine settings, in order to achieve the static equilibrium for the structure. Analysing several photos of such structures, the idea to model the structures at loop level as a truss framework, which is presented in [8], can be proposed. Figure 6.1 represents such a structure with marked dots in the middle of the loop heads. For small deformations the underlapping yarns (close to horizontal direction) and the loop legs (close to vertical direction yarn pieces) adjust the positions of their centers without any major additional motion.

If it is assumed, that these yarns remain straight and the effects of bending rigidity are neglected, the underlappings and the loop heads can be replaced by truss elements connected with joints in the loop centers, as demonstrated in Fig. 6.2.

A complete set of two guide bar structures presented in Fig. 6.3a can then be modelled as a framework presented in Fig. 6.3b.

6.2 Related Works

The replacement of the loops with mesh for plain weft knitted structure, is reported in [2]. For this case, the mesh of trusses is based on another principle. Six trusses are set to build each loop in a way, that the connection places are the contact points between the yarns, as visualized in Fig. 6.4. The application of the model is demonstrated for the simulation of uniaxial tension deformation and for draping simulation. Draping simulation of textile structures, based on the application of trusses, is as well reported in [4] and other works. The application of this model for warp knitted

Y. Kyosev, *Topology-Based Modeling of Textile Structures and Their Joint Assemblies*, https://doi.org/10.1007/978-3-030-02541-0_6

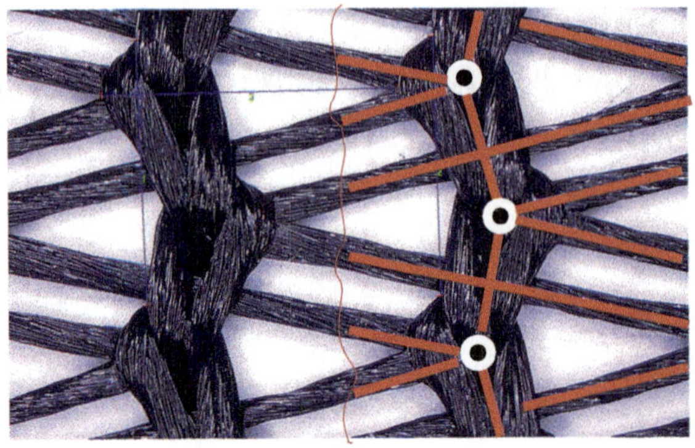

Fig. 6.1 Photography of warp knitted structure with the corresponding joint and trusses for its modelling

structures and for weft knitted structures with different patterns does not promise any advantages, because the topology of the connections between the yarns is varying with the variations of the patterns.

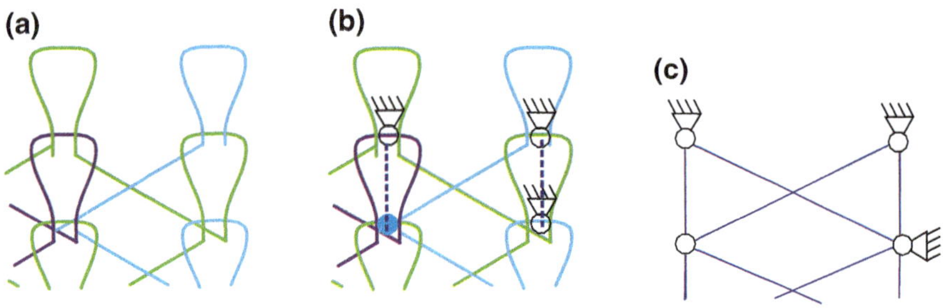

Fig. 6.2 Geometrical model of warp knitted structure and its corresponding truss framework equivalent. **a** loops **b** the loop head centers are taken as joints, each pair of legs is represented by one beam, **c** example of structure part with fixed boundaries

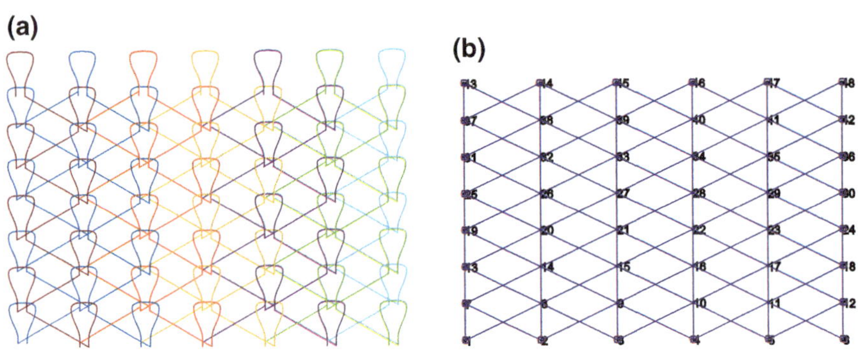

Fig. 6.3 Warp knitted structure produced by two guide bars (tricot-tricot) and its automatically generated truss based FEM model

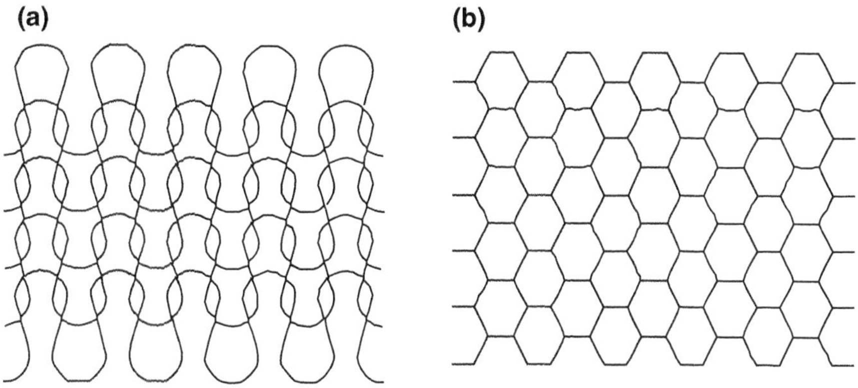

Fig. 6.4 Plain weft knitted structure and a hexagonal mesh of trusses used in [2]

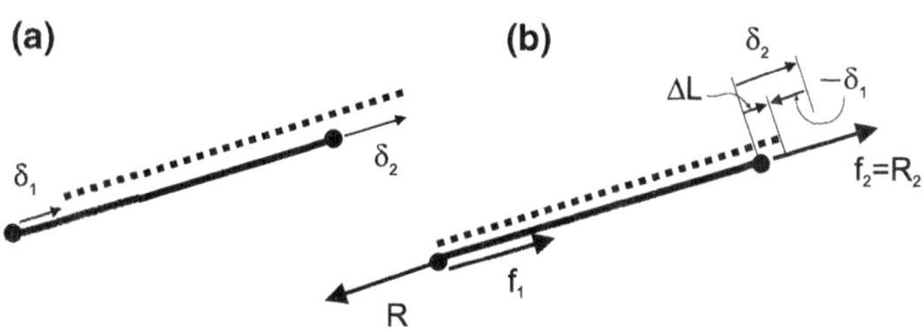

Fig. 6.5 Elongation of truss element, expressed in a local coordinate system

6.3 Theoretical Background of the FEM for Truss Based Frames

The reaction force in each truss is proportional to the relative elongation

$$R = \frac{EA}{L} \cdot \Delta L. \tag{6.1}$$

The nodal forces, caused by the elongation (or axial compression) $\Delta L = \delta_2 - \delta_1$ of the truss have the same magnitude as the reaction force, but with different sign, because of the different directions Fig. 6.5. These can be expressed using the displacements of the nodes as follows

$$\mathrm{f}_1 = \frac{EA}{L} \cdot (\delta_1 - \delta_2) \tag{6.2}$$

$$\mathrm{f}_2 = \frac{EA}{L} \cdot (\delta_2 - \delta_1) = \frac{EA}{L} \cdot (-\delta_1 + \delta_2) \tag{6.3}$$

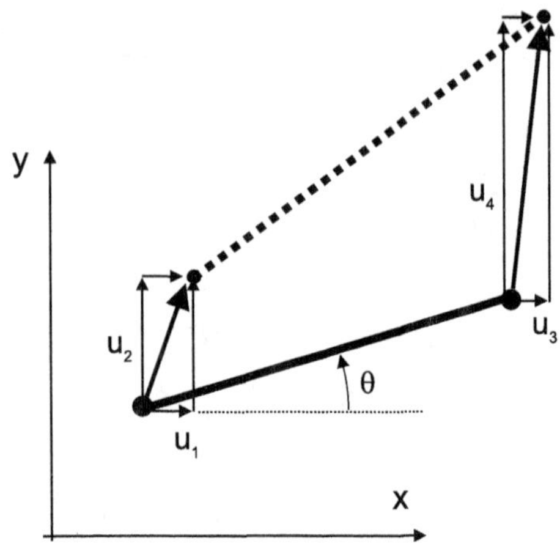

Fig. 6.6 Nodal displacements of truss element, expressed in a global coordinate system

These two equations in vector form can be rewritten as:

$$\begin{Bmatrix} f_1 \\ f_2 \end{Bmatrix} = \frac{AE}{L}\begin{bmatrix} 1 & -1 \\ -1 & 1 \end{bmatrix}\begin{Bmatrix} \delta_1 \\ \delta_2 \end{Bmatrix} \tag{6.4}$$

or shorten as

$$\mathbf{F} = \mathbf{K}_E \cdot \delta \tag{6.5}$$

where

$$K_E = \frac{AE}{L}\begin{bmatrix} 1 & -1 \\ -1 & 1 \end{bmatrix}. \tag{6.6}$$

Here K_E is the element stiffness matrix in a local coordinate system, the displacement vector in a local coordinate system is $\delta = \begin{Bmatrix} \delta_1 \\ \delta_2 \end{Bmatrix}$ and F is the force vector $\mathbf{F} = \begin{Bmatrix} f_1 \\ f_2 \end{Bmatrix}$ as well in local coordinate system.

Actually, each truss is oriented differently in the space and it is more efficient to use nodal displacements, expressed in global coordinates (see Fig. 6.6). In this case, each node has two degrees of freedom, so the total degrees of freedom of each truss are four, noted as a vector u:

$$\mathbf{u} = \begin{Bmatrix} u_1 \\ u_2 \\ u_3 \\ u_4 \end{Bmatrix} \tag{6.7}$$

The displacements in the element coordinate system can be expressed through the nodal displacements of the global coordinate system as

$$\delta_1 = u_1 \cos\theta + u_2 \sin\theta \tag{6.8}$$

$$\delta_2 = u_2 \cos\theta + u_2 \sin\theta \tag{6.9}$$

Which, in matrix form is

$$\delta = \mathbf{Tu} \tag{6.10}$$

where T is the transformation matrix.

$$\mathbf{T} = \begin{bmatrix} \cos\theta & \sin\theta & 0 & 0 \\ 0 & 0 & \cos\theta & \sin\theta \end{bmatrix} \tag{6.11}$$

The strain energy of the truss element can be written as

$$U = \frac{EA}{L} \int\limits_0^{\Delta x} \delta d\delta \tag{6.12}$$

in matrix form and replacing the displacement from local to global coordinate system

$$U = \frac{1}{2}\delta^T \mathbf{K_E} \delta = \frac{1}{2} u^T T^T \mathbf{K_E} T u = \frac{1}{2} u^T \mathrm{K} u \tag{6.13}$$

where

$$K = T^T \mathbf{K_E} T \tag{6.14}$$

is the stiffness matrix in the global coordinates.

The relation between the nodal forces and displacements in global coordinates then will be

$$\mathbf{F} = \mathbf{Ku} \tag{6.15}$$

which is the main equation used in the implicite Finite Element method.

For a complete mechanical system the displacement vector will consist of the displacements of all nodes, the forces vector—of all nodal forces and the stiffness matrix are build by adding the values of the element matrices to the previous values of the other elements on the corresponding positions. The algorithm for this, including the ways for consider the boundary conditions can be found for instance in [3] or several courses and books about programming of Finite Element Method.

The advantageof such a representation is that this equation presents a system of linear equations, which can be solved for a given stiffness matrix, based on the orientations and forces F. This can be done using well developed methods like the LU decomposition, available with almost all numerical analysis software packages. Mathematically it can be solved as

$$\mathbf{u} = \mathbf{F} \cdot inv\left(\mathbf{K}\right) \tag{6.16}$$

6.4 Implementation

The practical implementation of the loop based model requires several steps, which can be performed fully automatically within a software:

1. Topology: Create the topological based model of the structure, keeping information about the loop centers and underlappings (see Chap. 5)
2. Nodes: Create node at each loop center
3. Elements—Underlappings: Go through each yarn and create a truss between the loop centers, assigning the elasticity parameters of the yarn to the truss
4. Elements—Loop Head Legs: Go through each yarn and create a truss between the loop center and the one above; here two yarn pieces (legs) are building one truss, so the element should become assigned a doubled thickness
5. Boundary conditions—set boundary contagions on the boundary nodes
6. Build element and global stiffness matrices
7. Adjust the equation with the boundary conditions
8. Solve the system against the displacements
9. Update the coordinates of the loop centers
10. Update the yarn coordinates, considering the adjusted loop centers.

6.5 Boundary Conditions

Depending on the type of the tasks, different boundary conditions can be set. As the geometrical algorithms for the creation of 3D geometry of warp knitted structures generate the loops vertically oriented, but in the reality they are rotated to some angle, the intended application of the truss framework model is to be used for performing relaxation of the structure and to achieve a more realistic model. For this purpose two different approaches with the boundary conditions are possible:

- periodic boundary conditions for one repeat
- fixed two or four sides of a larger sample.

From the computational point of view, using the periodic boundary conditions is a very efficient way, but its implementation is connected with a complex procedure in concern to their definition. The warp knitted structures usually have larger repeats and have several yarns going through the unit cell. For the preparation of the periodic boundary conditions, the intersection point between the yarn and the unit cell boundary of each yarn has to be calculated and at its place an additional node have to be generated. All nodes from the left hand side of the unit cell have their sibling on the right hand side of the unit cell (Fig. 6.7b). Each pair of siblings has the same value of the horizontal displacements, but with opposite value (symmetric) and identical vertical displacement (Fig. 6.7c).

The alternative approach is to model the larger part of the structure and fix for instance all four boundaries or the two of their sides, as presented in the mesh of

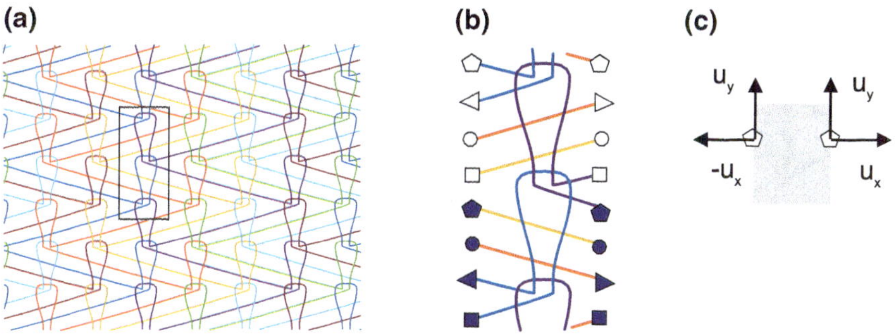

Fig. 6.7 The larger sizes of the warp knitted structures force the researchers to use only their unit cells for computations. **a** presents a marked unit cell of such structure; **b** all yarns, which intersect the unit cell boundary have to be split into two pieces and special nodes at the breaking points are created; **c** each node—couple has symmetrical x displacements and identical y displacements

another structure in Fig. 6.9. In this case, there is no special procedure required for splitting trusses and the size of the model is or can be significantly larger. Because such a model or even larger sized models are no problem for modern computers, this approach was preferred in this case, due to its simplicity.

6.6 Implementation and Verification

The implementation of this algorithm, used for the calculations in [5, 6, 8] and later papers, is realized as a set of MATLAB scripts and functions. This allows the complete procedure to be executed, from reading the sample file with coding and all machine and yarn parameters, creation of the geometrical model with the complete access to all intermediate data for its generation, creation of the elements and nodes and then recreation of the yarn geometry to be integrated smoothly, without any need of interfaces between the single parts.

Figure 6.8a presents a single guide bar structure generated using a geometrical model, refined using the truss framework system Fig. 6.8b and a photo of a real structure of it Fig. 6.8c. The changed orientation of the loop heads in the middle picture demonstrates the optical similarity to the real one and proofs the suitability of this method for refinements of such structures (Fig. 6.9).

6.7 Possible Extension for Double Needle Bed Structures

The model can be extended and applied for spacer structures as the one presented in Fig. 6.11, which is produced on double needle bed machines. In such a case, truss elements in the 3D space have to be used, so each receives three degrees of freedom

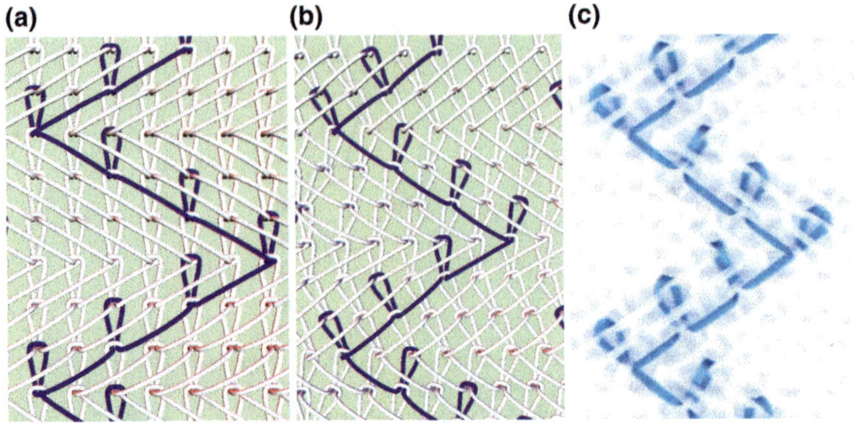

Fig. 6.8 Validation of the truss framework method. **a** Model using geometrical algorithms (loops are vertical); **b** after application of the truss framework model—loop heads are oriented to left or right side similar to the real structure, presented in (**c**)

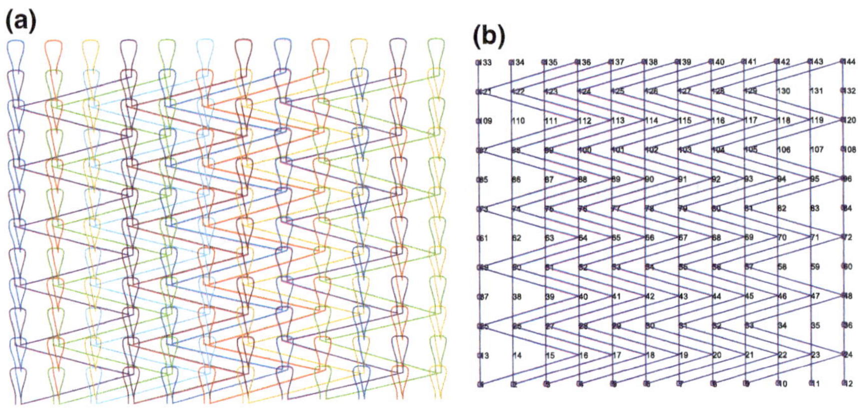

Fig. 6.9 **a** Larger structure produced with two guide bars. One of these produces lock-stitch, (series of loops in same vertical line)so that the boundary areas always have loops, being independent from the fact, if the other guide is currently laying yarns on the side needles or not, **b** presents the FEM model of this structure [6]

in the global coordinate system. For several applications, for which the draping of the structure and its macro behaviour is investigated, the system can be efficient enough.

As the structure with all nodes, elements and boundary conditions (Fig. 6.10) is stored as data within the program, it is no problem to write this in a format, suitable for some of the commercial or open source FEM. In concern to some packages the user has more freedom regarding the adjustment of the boundary conditions, load cases, solvers and they can be integrated within larger simulations for instance for the calculation of the behaviour with another contact body, which having a complex geometry. The repetition of the structure in the vertical direction is longer and consists

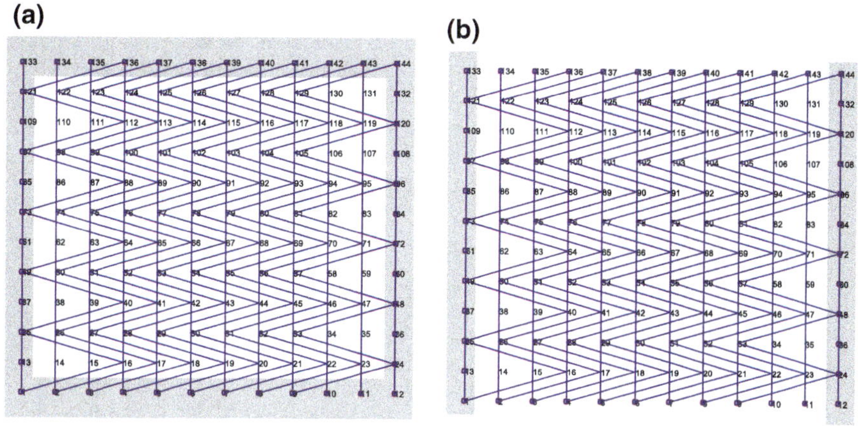

Fig. 6.10 The FEM model of the structure presented in Fig. 6.9, **a** with all boundaries fixed as on biaxial tester or on some special stentering frames; **b** only fixed vertical sides

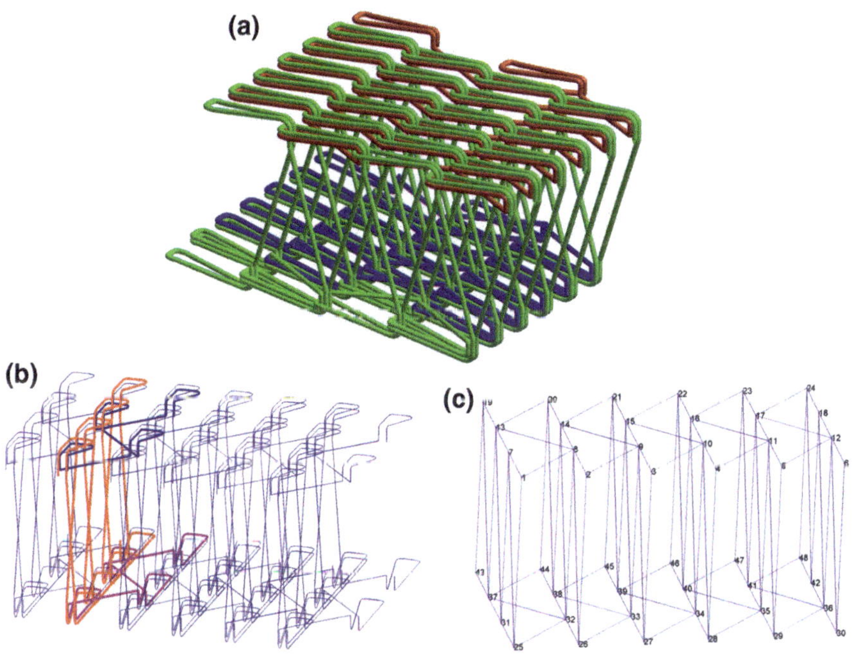

Fig. 6.11 Geometrical model of simulated spacer structure

of alternating connections between one column and the left and right one. The build net structure is open and drapeable.

Figure 6.12a shows a small part of a mesh type double needle bed warp knitted spacer structure and the mesh of it, imported in ANSYS(R) (Fig. 6.12b). A few steps

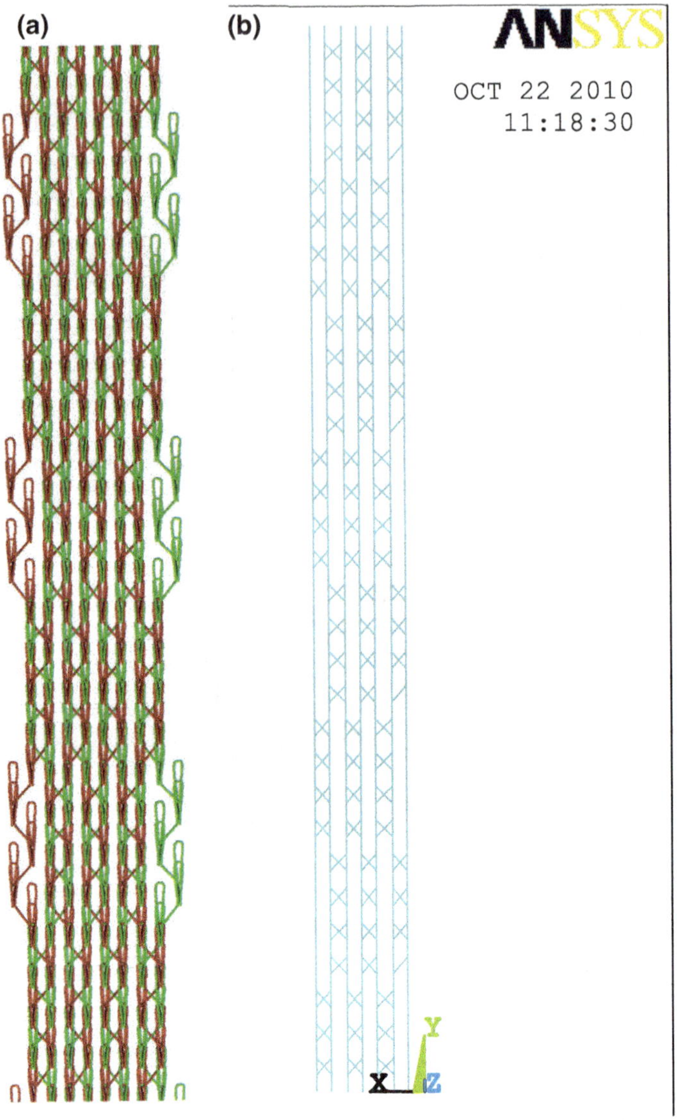

Fig. 6.12 **a** Mesh type double needle bed warp knitted spacer structure and **b** FEM mesh of this structure imported in ANSYS(R)

of the horizontal extension simulation with ANSYS are visible in Fig. 6.13. The usage of an external FEM pre-processor, solver and post-processor gives additional user friendly possibilities for adjusting the model and adds simulation options, but does not reduce the limitations of the model, based on its assumptions.

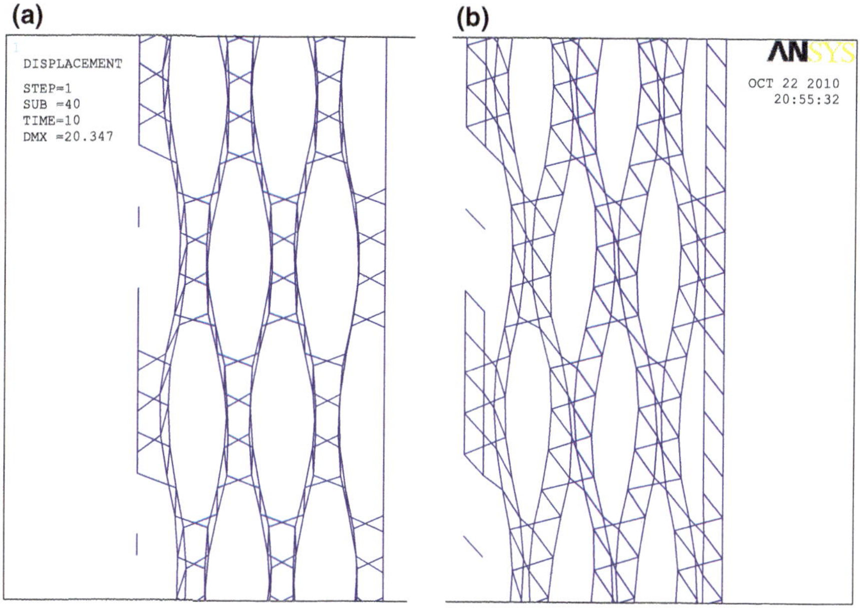

Fig. 6.13 Two steps of the horizontal extension simulation of the structure from Fig. 6.12 within the software ANSYS

6.8 Limitations of the Method

6.8.1 Stiffness Matrix and Singularities

The global stiffness matrix for such system (with smaller size, for better visibility) is with well presented main diagonal and with small band width, as visualized on the Fig. 6.14.

Normally, linear systems with such matrices can be solved using direct methods. During the testing it was found out, that some changes of the warp knitting pattern, the threading or the boundary conditions lead to badly scaled matrix and invalid results.

Looking back at the FEM mesh it can be seen, that even the warp knitted structure causes the creation of triangular like cells of trusses, these triangles are build by four! trusses and represent deformed tetragons. In the vertical direction, usually two trusses build one line, which is connected with another two trusses. Therefore, the optically stable looking, statically determined triangle, is actually a kinematic mechanism (planar four-bar linkage), which is underdetermined and has an additional degree of freedom (Fig. 6.15).

Considering only one element in Fig. 6.15c, having $t = 4$ trusses (links); $r = 0$ reactions (removed degrees of freedom in the lagers) and 4 joints (each with 1

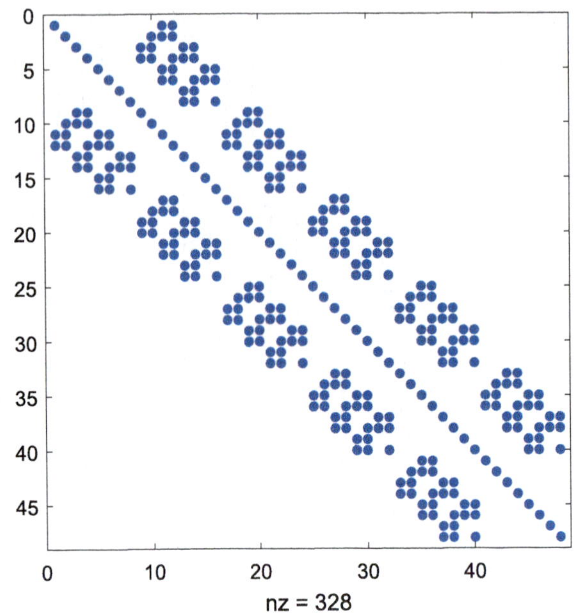

Fig. 6.14 Places, where the global stiffness matrix has non zero elements

degree of freedom—only rotation) , the mobility formula according the Chebychev-Grübler-Kurzbach criterium [1] degrees of freedom of such an element is

$$M = 3 \cdot (N - 1 - j) + \sum_{i=1}^{j} f_i = 3 \cdot (4 - 1 - 4) + 4 = 1 \tag{6.17}$$

Degree of freedom 1 means, that the complete linkage can perform one motion—in this case rotation. From the point of view of unknown variables and equations, the one "element" has $j = 4$ joints, with 2 degrees of freedom (for the planar case)—the motions in x and y directions, which results to $2 \cdot j = 2 \cdot 4 = 8$ unknown displacements. Each link adds one equation as constrain between the joints, so there are t constraints for the lengths of the links. Each joint provides two equilibrium equations (sum all forces in the x and y direction). Finally there are $t + 2j = 4 + 2 * 4 = 12$ equations and $2 \cdot j$ unknowns or

$$f = t + 2 \cdot j - 2 \cdot j = 4 \tag{6.18}$$

equations more than unknowns. In order to solve such problem, four degrees of freedom of the system has to be constrained by fixing carefully selected joints. This fixing adds four additional unknown variables—the unknown reactions at the fixed places and the system could becomes well determined and solvable as normal linear system of equations.

Analysis of the structure presented in Fig. 6.9 shows, that its model consists of:

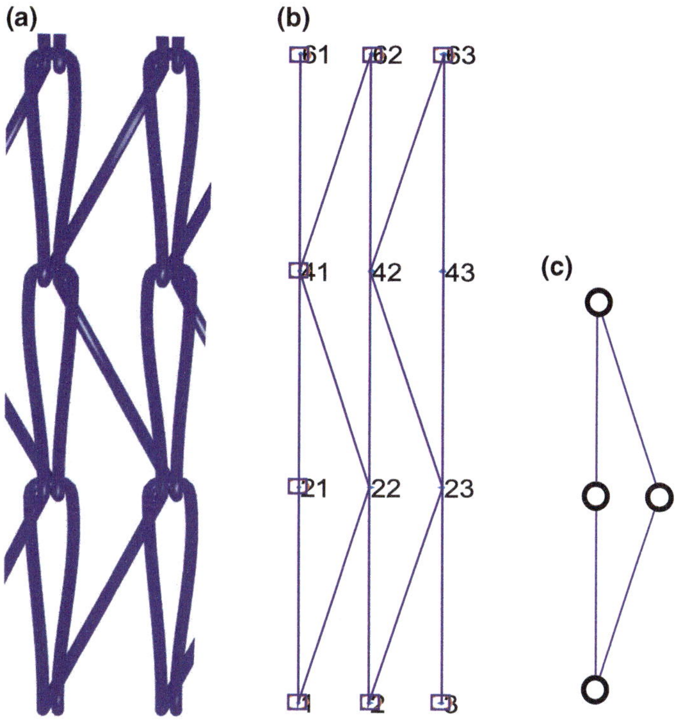

Fig. 6.15 **a** Elements of structure, produced with one guide bar, **b** corresponding FEM mesh, **c** one group of trusses

- $t = 11 \cdot 12 + 9 \cdot 11 = 231$ trusses
- $2 \cdot j = 2 \cdot 12 \cdot 12 = 288$ degrees of freedoms in the joints

so following

$$f = 231 + r - 288 = r - 57. \tag{6.19}$$

In order to get the system statically determined, which means $f = 0$, exactly $r = 57$, well selected lager reactions should be added in order to keep the system determined, or in the meaning of mathematics—to have the same number of unknowns as the number of the equations, if force equilibrium at each node is applied.

If all boundary nodes become fixed and are then removed, they will be $r = 44 \cdot 2 = 88 > 57$ so the system becomes overdetermined, but stable. Mechanically, this can simply be recognized by considering the starting triangles on the left and right hand side (as their two vertical trusses are completely fixed, so they disappear for the system).

If only the left and right hand side of the nodes receive both fixed degrees of freedom, then not enough degrees of freedom are removed, $r = 2 \cdot 12 \cdot 2 = 48 < 57$ which leads to 9 degrees of freedom not being removed and therefore, the system can not be solved directly.

This simple example shows, that the model, based on the solution of linear system and static analysis, is very sensitive. Of course, its determinacy depends on the boundary conditions—the number of the fixed joints, but the internal structure—the number of the trusses, depends on the pattern of the warp knitting structure. As the textile engineers change such a pattern frequently, it is not possible to apply it in its simplest form as an universal solution method. Its implementation should be connected with a conditional selection of a solver:

- If the system is determined, a linear system solver can be used. Such a system is fast and offers very quick solutions
- If the system is overdetermined, an interactive method (least squares) should be used to find the solution
- If the system is underdetermined, it can have infinite number of solutions, if any. In this case additional constraints has to be added.

6.8.2 Model Limitations

The truss framework was defined for yarns, which build loops. Additional structural components of the warp knitted structures are the weft yarns, these do not build loops, they are connected between the loops. The integration of the weft yarns looks unproblematic at the beginning, but only if the yarns are thin. Figure 6.16 shows two such structures, in which the weft yarns can be modelled as attachements to the loop heads in the same matter as the normal underlappings. Once the structure becomes more compact or the weft yarns are thicker, as presented in Fig. 6.17, the side connection to the loop heads remains valid, but the mobility of the other loops becomes limited. Such thick weft yarns, as used for glass and carbon fiber reinforced composites, normally dominate the weight in the structures and as well determine the orientations of all loops and yarns with which they are connected. The inclusion

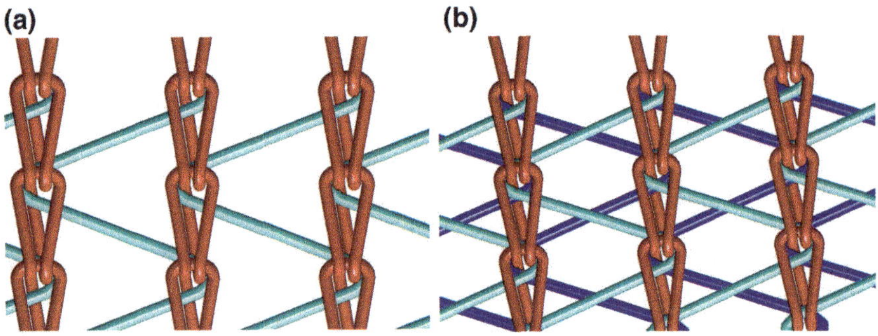

Fig. 6.16 **a** Idealized model of structure with loops and weft yarns; the weft yarns will cause movenet of the loop centers to one or other side, **b** the same structure but with two weft yarn groups, keeping the loops in equilibrium state vertically

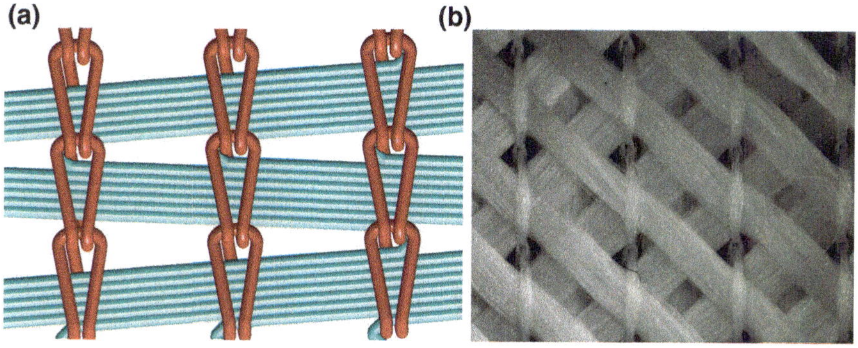

Fig. 6.17 Model of structure with multiple or thicker weft yarns and thin loops, **a** model of weft with partial weft placed from one guide, **b** photo of structure with two full weft (magazin) systems

of such yarns in the truss framework model would require additional contact points with the loops itself, because they often go trough several loops. Actually, when such thick weft yarns are used (as the reinforced weft knitted structures from the previous chapter, Sect. 5.10), the loops are only used to keep the weft yarns together and the modelling of the loop orientation loose their purpose in such structures.

Because in warp knitting, each guide can produce loops and wefts insertions, depending on the coding, the availability of the weft yarns has to be checked for each structure in order to decide, if the truss framework model can be used or not.

6.8.3 Elasticity Module of the Trusses

An interesting question is the identification of the elastic properties of the trusses in the truss framework model. For the initial calculations above, the elasticity modulus of the yarns was applied in the trusses. Actually, the loops build by curved yarns. Therefore, during the axial tension load of one loop reacts initially with a force, which is depending on the unbending forces of the curved parts of the loops and it does not depend directly on the elasticity modulus of the yarns—see Fig. 6.18. The yarn elasticity modulus becomes important after the loops are stretched to a certain degree. But this means, that the trusses should be described with a nonlinear material model, having one, lower initial elasticity modulus E_1 for lower deformations and a second, higher elasticity modulus E_2 for larger deformations, or generally

$$EA = EA(\varepsilon) \tag{6.20}$$

and following the stiffness matrix of the system becomes as well nonlinear

$$K = K(\varepsilon) \tag{6.21}$$

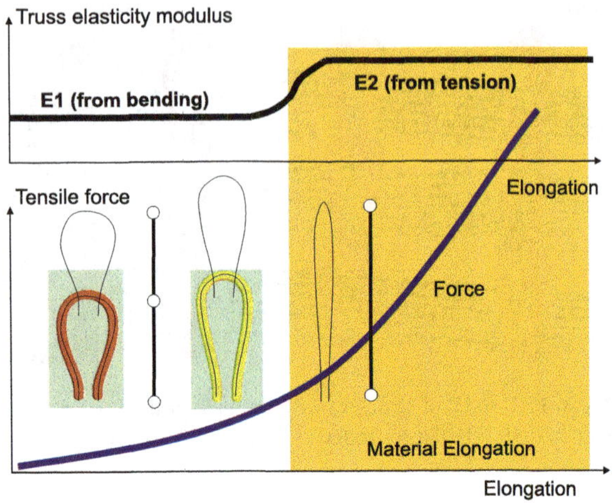

Fig. 6.18 The elasticity modulus of the trusses depends on the type of the loops and the bending stiffness of the yarns. In the initial area the tensile response of the loops, which would determine the truss elasticity modulus, depends on the bending stiffness of the yarns, and after the loop is straight, then the tensile elasticity modulus of the yarns becomes important for the elasticity of the truss element [7]

For modelling this, the system (6.15) has to be solved as nonlinear system of equations.

Another limitation of the model is, that the loop center is assumed to be the only moving part and the yarns within the loops do not redistribute their lengths. This is valid only for small to medium deformations of the structures, but the limit is depending on the pattern. Once the yarns start sliding and redistributing their lengths between the loop heads and the underlaps (Fig. 6.19) the validity of the truss framework model is at its end.

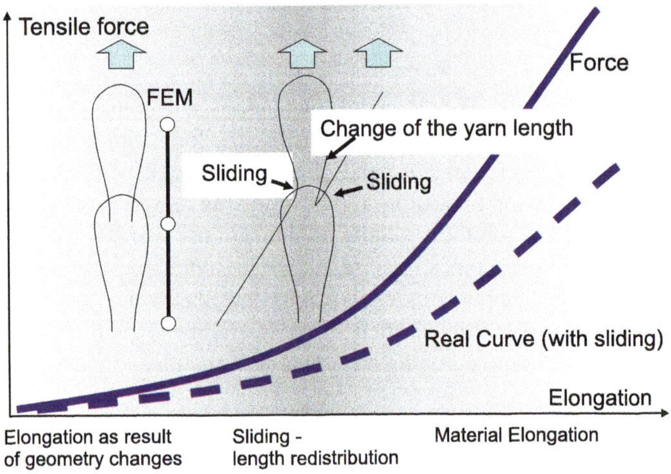

Fig. 6.19 The truss framework model assumes no slippage in the loop heads. Once the deformations of the structure becomes larger, the slippage can occur and the calculated force-elongation diagram will be overestimated

6.9 Conclusions

The developed truss framework model for warp knitted structures allows the automated program based generation of FEM mesh, based on the geometrical model of the structure, without the need of human intervention. Under the assumptions that there is no sliding between the yarns and that there are no thick weft yarns in the structure, the model can be used for the relaxation procedure of warp knitted structures in order to refine the originally geometrically generated geometries. Other suitable applications are the draping, tensile and other mechanical tests simulations in the area of small deformations. In both cases, the application of the truss framework model is connected with minimal computation resources. Due to the dependence of the stability of the system on the pattern, the implementation of this model in productional software requires checks of the degrees of freedom of the system. It can not be used as an exclusive computational model in software. For the cases, within which the assumptions are not fulfilled, another algorithm should be available.

References

1. Angeles, J., Truesdell, C.: Rational Kinematics. Springer, Berlin (1989)
2. Araújo, M., Fangueiro, R., Hong, H.: Modelling and simulation of the mechanical behaviour of weft-knitted fabrics for technical applications. Part III: 2d hexagonal FEA model with non-linear truss elements. Autex Res. J. **4**(1), 25–32 (2004)
3. Belytschko, T., Liu, W., Moran, B., Elkhodary, K.: Nonlinear Finite Elements for Continua and Structures, 2nd edn. Wiley, Chichester (2014)
4. Cherif, C.: Drapierbarkeitssimulation von Verstärkungstextilien für den Einsatz in Faserverbundkunststoffen mit der Finite-Elemente-Methode, als ms. gedr. edn. Shaker, Aachen (1998). http://www.worldcat.org/oclc/722528088
5. Kyosev, Y.K., Renkens, W.: Computational mechanics of knitted structures—critical overview. In: Eberhardsteiner, J. et al. (ed.) European Congress on Computational Methods in Applied Sciences and Engineering (ECCOMAS 2012) (2012)
6. Kyosev, Y.K., Renkens, W.: Rechnergestützte mechanik von maschenwaren - eine kritische übersicht: Ausgabe 1/2-2013. Textilplus **1**(1/2) (2013)
7. Kyosev, Y.K., Renkens, W.: Simulation of the mechanical behaviour of warp knitted fabrics (2013)
8. Renkens, W., Kyosev, Y.: 3D simulation of warp knitted structures—new chance for researchers and educators. In: 7th International Conference TEXSCI'10 (2010)

Part IV
Woven Structures and Sewing Stitches

Chapter 7
Notes About Topological Methods for Woven Structures

7.1 Introduction

The woven structures and their modelling methods are very intensively investigated. The author of this book recognized the large number of unexplored fields in the modelling of knitted and braided structures and concentrated mainly in these areas in the past years. This book try to bring the modelling methods of these (knitted and braided) structures at the level, at which is the modelling of the woven structures nowadays. Therefore, the woven structures are not objective of the current work and this chapter only has the task to give basic information and is presented for the sake of completeness of this work. Still, some important issues related to the coding of the woven structures are discussed.

7.2 Single Layer Structures

The woven fabrics belong to the most commonly used and investigated textile structures. The orthogonality between the two yarn systems—weft and warp yarns—is significantly simplifying for the modelling process. The coding, used for woven structures is based on mesh with the crossing points of the warp and weft yarns (Fig. 7.1). The warp yarns can be located at a upper or bottom position in the weaving shed. Their position is controlled by the held frames, and the controlling code for the frames (up or down) is additionally used for the topological description of the woven structure. If the warp yarn is up, the square is filled, if the warp yarn is down, it is empty. This coding (converted to 1 for up and 0 for down) is used for the control of the held frames or the Jacquard unit of the weaving machine and also for the reconstruction of the geometry of the woven fabric. The weft yarns remain in the inverted position of the warp yarns. Based on this coding, the topology of the single layer woven structures is created. Usually, researchers go even further and precisely describe the yarn orientation using geometrical models (not only topological). There

Y. Kyosev, *Topology-Based Modeling of Textile Structures and Their Joint Assemblies*, https://doi.org/10.1007/978-3-030-02541-0_7

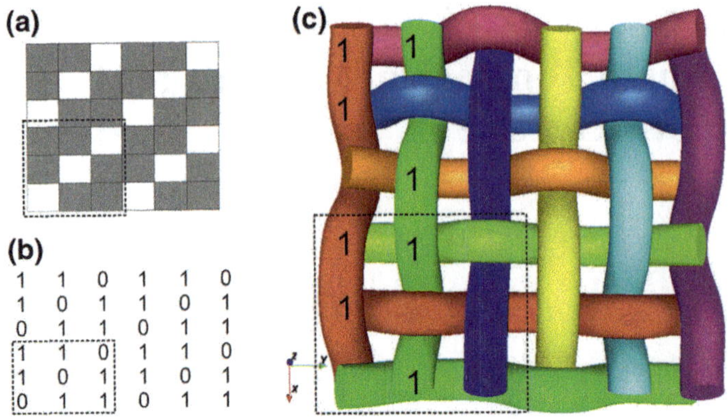

Fig. 7.1 Coding of woven structures and its simulation: **a** pattern with the marked repeat, **b** the same pattern in matrix form, **c** simulated with the software TexGen [18] woven structure, corresponding to this pattern. Reprinted from D. Veit (Ed.) "Simulation in Textile Technology", Y. Kyosev, "Simulation of wound packages, woven, braided and knitted structures", pp. 266–314, Copyright (2012) with permission from Elsevier [4]

are more than 200 research papers, covering different aspects about the geometrical modelling and the properties of woven structures.

One of the first works in this field is the one of Peirce [10] from 1937 about a woven structure, based on a circular cross-section of yarns and about the modelling of yarn paths using circular arcs and lines. This model is used for the creation of graphical relations between the parameters, used for most of the structures by Love [9]. Detailed investigations about the weavability, thickness, fabric degree of tightness etc. of woven fabrics are reported in the series of papers of Seyam [12–14].

This model is extended in several papers for different shapes of the yarn cross-section (lenticular, eliptic) and for different kinds of approximations of the yarn paths (eliptic, parabolic, cubic, natural splines etc.). Experience shows that each of these models can be valid for certain cases and is not applicable to other cases. For plain fabrics, such a description can be found in [1, 11]. A very precise implementation of the different models in a software, which chooses the suitable yarn path approximation function depending on the yarn cross-section form is the program Wisetex [5–7]. The software is proven in a large set of investigations as for instance [3, 8] etc.

7.3 Multiple Layer Structures

For multiple layer woven structures the coding is extended to numbers, which define at which horizontal level the warp yarn is placed for the current crossing point (Fig. 7.2). In the standard case it is assumed, that there are weft yarns at each of the levels.

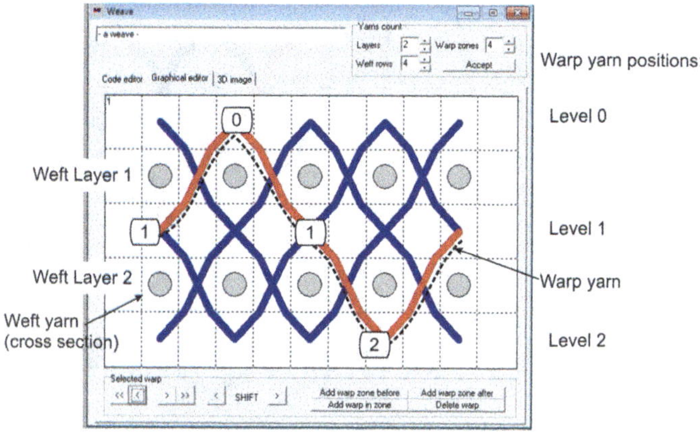

Fig. 7.2 Coding of multilayer woven structures—the number in this case shows at which level the warp yarns is placed. Example based on Wisetex software [3, 5]

The topology of the structure is implemented in both, Wisetex and in the weaving wizards of TexGen software [15, 18], but Wisetex uses geometrical (meaning one level higher accuracy) models, while the Texgen is based on the topological representation, based on the key point coordinates only. Which kinds of models are integrated into commercial design softwares, like EAT GmbH "The DesignScope Company" [2], is not reported in the literature, but looking at the figures the implementation seems to be based on the topological representation.

Some special algorithms for hollow or angle interlock structures, for which the simulation of the structure is based on the coding of the machine, are implemented by TexEng LTD [16].

7.4 Irregular Pattern

The topology of the woven structure describes the relation between the positions of the yarns in the 3D space very well, but for the designer this is often not enough. In many cases, the visualization of the differences in the connection between multiple layers is important. Especially interesting for the producers is the automatic recognition and visualization of hollow areas, as reported in [17] and possible in [2]. There are a lot of multiple layers of woven structures, in which the wefts are not inserted regularly at each level, but only at certain places. In this case their position in the vertical direction and horizontal direction is not always defined or not easily recognized. Some of these structures can be represented by shifting a single layer as it is possible in Wisetex (Fig. 7.3). From the topological point of view, the interlacement remains the same, but the geometrical configuration of the structure is different. Generally, with the increased use of computer controlled drives in the modern

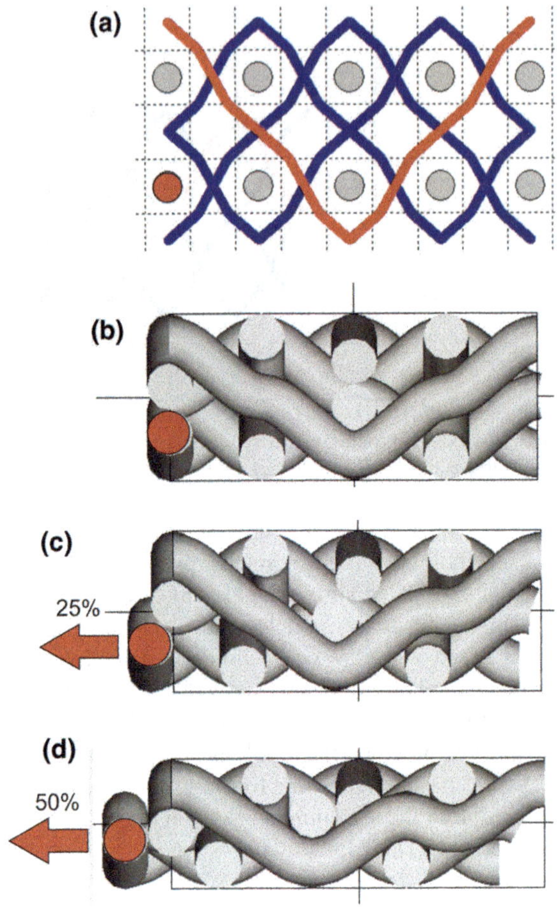

Fig. 7.3 Shifting of layer in Wisetex as possibility for adjustment the position of the weft yarns. **a** Topological definition, **b** normal geometry, **c** the bottom layer shift at 25% from one step size, **d** bottom layer shift at 50% from one step size

weaving machines, more and more structures will be produced, where the crossing points between the weft and warp yarns are not distributed on a fixed squared mesh, but at different places in the plane. For the correct representation of such structures the information about the take off speed (which define the placement of the weft yarn) and in some more complex cases also the yarn tension is required additionally to the topological record (the coding).

The weavers use a non-numerical way of coding of multilayer structures. They use set of signs and each sign has some meaning, which represents an important topological information. This information is more than the information of a single number, giving the position. This is demonstrated in Fig. 7.4—both signs—"dot in a square" and the "filled square" are numerically coded by 1. During the weaving the warp yarn is in the upper position, but the warp yarn, which is set to the upper position with the "dot sign" is not participating in the interlacement with the current weft and remains building upper layers, while the yarns, set to the upper position by

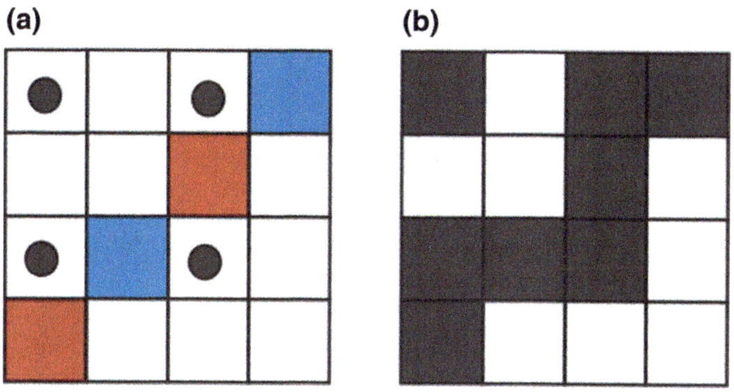

Fig. 7.4 The weavers use a non-numerical way of coding of multilayer structures. They use set of signs, each of these represents important topological information

"filled square", participate in the interlacement. This is already a lot of information on how the yarns are connected in the space. The usage of colors additionally allows to understand and keep in mind which yarn are participating—for instance the red squares in the Fig. 7.4a are used for the warp yarns, building the upper layer and the blue squares, presenting the warp yarns, are building the bottom layer. For the control of the machine all this information is not important. The machine uses binary coding for the frames up or down—Fig. 7.4b and the structure is built based on this motion and the forces between the yarns. For modelling algorithms actually such information could become important for the reduction of the amount of calculations in order to achieve the result. This is a question of choice, because if the weaver is thinking and coding completely, the computer has an easy job. If the human does not understand the process of weaving (as it becomes even more and more modern), the scientists have to find a way to implement algorithms for finding the proper configuration, using intensive FEM simulation, energy minimization etc.

7.5 Conclusions

This small chapter includes some remarks and points out some main references, related to the topological modelling of the woven structures. The principle of the coding of woven interlacement for single layer and multiple layer structures is explained. Some open issues about the modelling and coding of irregular structures are mentioned.

References

1. Behera, B., Hari, B.K.: Woven Textile Structure: Theory and Application. Woodhead Publishing Ltd, Cambridge (2010). http://www.worldcat.org/oclc/610819829
2. EAT: Designscope (2011). http://www.designscopecompany.com
3. Kyosev, Y., Lomov, S., Küster, K.: The numerical prediction of the tensile behaviour of multilayer woven tapes made by multifilament yarns. In: Kyosev, Y. (ed.) Recent Developments in Braiding and Narrow Weaving, pp. 69–80. Springer International Publishing, Cham (2016). https://doi.org/10.1007/978-3-319-29932-7_7
4. Kyosev, Y.K.: Simulation of wound packages, woven, braided and knitted structures. In: Veit, D. (ed.) Simulation in Textile Technology, pp. 266–309. Woodhead Publishing Limited (2012). https://doi.org/10.1533/9780857097088.266
5. Lomov, S.: Integrated textile preprocessor WiseTex, ver.3.2—computational models, methods and algorithms. KU Leuven, Leuven (2013)
6. Lomov, S., Bernal, E., Koissin, V., Peeters, T.: Integrated textile preprocessor wisetex, version 2.5: computational models, methods and algorithms (2012)
7. Lomov, S., et al.: Wisetex (2011). http://www.mtm.kuleuven.be/Onderzoek/Composites/Research/meso-macro/textile_composites_map/textile_modelling/textile_modelling_fe
8. Lomov, S.V., Willems, A., Vandepitte D., Verpoest, I.: Simulations of shear and tension of glass/pp woven fabrics. In: Juster, N., Rosochowski, A. (eds.) The 9th International Conference on Material Forming ESAFORM, pp. 783–786 (2006)
9. Love, L.: Graphical relationships in cloth geometry for plain, twill, and sateen weaves. Text. Res. J. **24**(12), 1073–1083 (1954). http://trj.sagepub.com/cgi/content/abstract/24/12/1073
10. Peirce, F.T.: 5–the geometry of cloth structure. J. Text. Inst. Trans. **28**(3), T45–T96 (1937). https://doi.org/10.1080/19447023708658809. http://www.tandfonline.com/doi/abs/10.1080/19447023708658809
11. Postle, R., Carnaby, G., de Jong, S.: The Mechanics of Wool Structures. Ellis Horwood Limited (1988)
12. Seyam, A., El-Shiekh, A.: Mechanics of woven fabrics: part i: theoretical investigation of weavability limit of yarns with thickness variation. Text. Res. J. **60**(7), 389–404 (1990). https://doi.org/10.1177/004051759006000704
13. Seyam, A., El-Shiekh, A.: Mechanics of woven fabrics. Text. Res. J. **63**(7), 371–378 (1993). http://trj.sagepub.com/cgi/content/abstract/63/7/371
14. Seyam, A., El-Shiekh, A.: Mechanics of woven fabrics. Text. Res. J. **64**(11), 653–662 (1994). http://trj.sagepub.com/cgi/content/abstract/64/11/653
15. Sherburn, M.: Geometric and mechanical modelling of textiles. Ph.d, University of Nottingham, Notthingham, July 2007. http://etheses.nottingham.ac.uk/303/1/thesis-final.pdf
16. TexEng Software Ltd: Knitgeomodeller (2011). http://www.texeng.co.uk/knitgeomodeller.html
17. TexEng Software Ltd: Weave engineer (2011). http://www.texeng.co.uk/2dfabricsl.html
18. TexGen (2007). http://texgen.sourceforge.net/index.php/Main_Page

Chapter 8
Topology Based Modeling of Sewing Stitches

8.1 Introduction

In most cases, final textile products have to be assembled from a several separate parts by stitching, welding or other assembling method. Welding or melting partially destroy the yarn orientation at the contact place or are melt them into solid or flexible composite structure. The modelling of such places requires "multiphysics" modelling with methods that can cover processes of multiple phases, which is not goal of the current work. The most commonly used method for assembling textile parts is sewing. The behavior at the stitched place can be investigated at the yarn level, if additionally to the structure also the geometry of the stitches is available. This chapter presents the rules for modelling the main stitches used in the textile production. The method and the issues, required for the last step, the stitching or assembling, are subject of the next chapter.

8.2 State of the Art

There are various applications in which the properties of the sewing stitches significantly influence the properties of the final product. Examples of such cases are parachute fabrics [10] for which both, the orientation of the fabrics and the stitches, are important and their improper selection can costs human lives. Stitching and sewing show an advanced development in the area of composites, because they allow efficient production of complex 3D forms [6, 9]. The stitches change the compaction of the fabrics [2, 7, 11] and in this way change the local fibre volume fraction and the complete properties of the composite parts. Both, the thread tension on the sewing machine [4] and the stitch arrangement [3], are found to be important parameters. All these investigations demonstrate that the modelling of the geometry of sewing threads and their interaction with the textile structures is important. The possibility for creating geometry of the complete assembly including the geometry of the fabrics

Y. Kyosev, *Topology-Based Modeling of Textile Structures and Their Joint Assemblies*, https://doi.org/10.1007/978-3-030-02541-0_8

and the sewing yarn will speed up the numerical simulations and optimization of all kind of joint structures.

The main sewing stitches are actually no complex structures and can be modelled using simplified geometrical rules, and there are not many research papers available about these. One example is the Bachelor Thesis of Nora Brinkert, co-supervised by the author [1], which succeeded to document the rules for the coordinates of the main points of the main stitches and to visualize these using TexGen [5, 12]. This chapter presents some specific rules for the calculation of the coordinates of the yarns in sewing stitches, based on the experience of the author. The stitch classes are following the ASTM D-6193 and ISO 4915:1991 norms. Explanations on how to build these stitches and their properties can be found for instance in [8].

8.3 Hand Stitch—Class 209

The simplest and the oldest stitch is the hand stitch, produced with one needle going through the textiles, known as class 209 and actually requiring complex machines for its production. The modelling is included in this abstract, because it presents the basic rules and problems of the stitch modelling. Still, it is used in several specific applications in the fashion area (decorating stitches) and it has the potential to be used in smart textiles, as the yarn is available from both sides of the material without any interlacement with another sewing yarn with high curvature places (Fig. 8.1). demonstrates the four most important points of such a stitch, marked as 1, 2, 3 and 4 and their coordinates in the local coordinate system, where x is the stitch direction, y is the direction of the needle motion or the fabric thickness, starting from the plate of the machine. The only parameters for this stitch are the stitch length t, the fabric thickness h and the yarn diameter, for which here the yarn radius r is used, due to simplification reasons (Figs. 8.2 and 8.3).

This model is sufficient represent the topologically correct way of the sewing stitch. Actually, if the stitched assembly has to be modelled at fiber level, the sewing yarn will build arcs through these keypoints and will interlace the yarns of the sewed fabric (Fig. 8.4, the line with 4 keypoints case (b)). To just use lines for connecting the key points leads to a clear stitch, but produces problems during the meshing of the

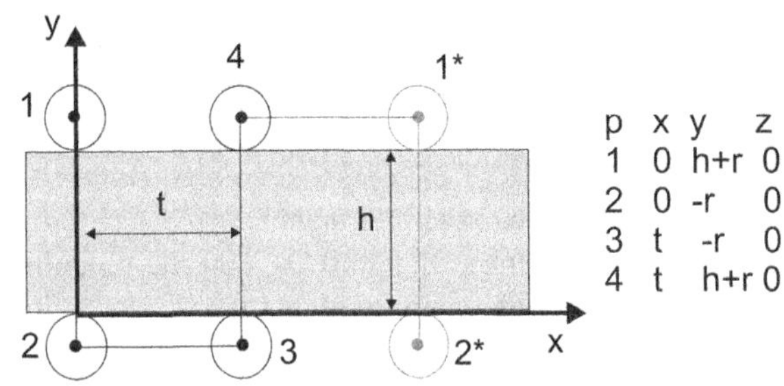

Fig. 8.1 Simplest selection of the keypoints for hand stitch (209)

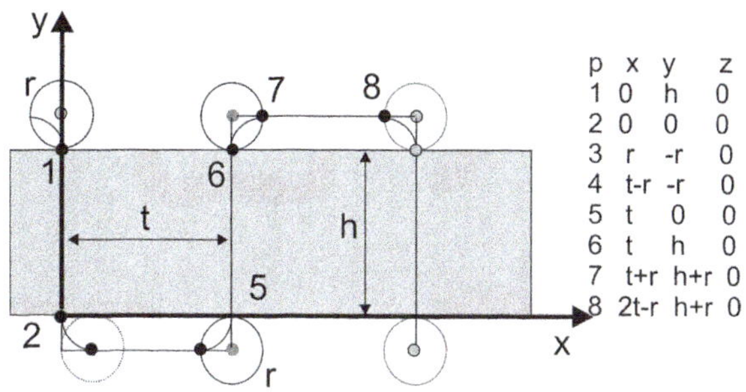

Fig. 8.2 Extended variant of keypoints for stitch 209 with two points at edge

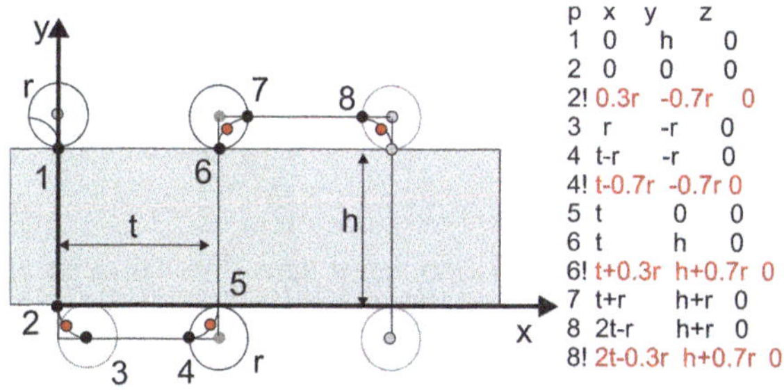

Fig. 8.3 Detailed variant of the key points for stitch 200 with 3 points at each arc

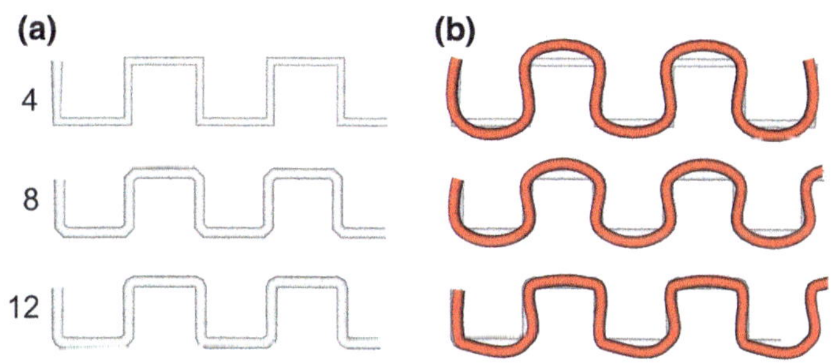

Fig. 8.4 Comparison of models of stitch 209 with different number of points, **a** without spline trough the keypoints, **b** with spline trough the keypoints

solid body for instance fore FEM models at the edges. Applying splines interpolation through the key points helps to avoid this problem, but the curve is positioned far around the ideal position.

In order for these problems to be reduced, the points where the yarn changes its curvature can be used as key points instead of the edge points. This doubles the number of the required points to 8 (Fig. 8.2). With 8 points both non-smoothed and the smoothed yarn volumes are looking better (Fig. 8.4, the line with 8 keypoints cases (a) and (b)).

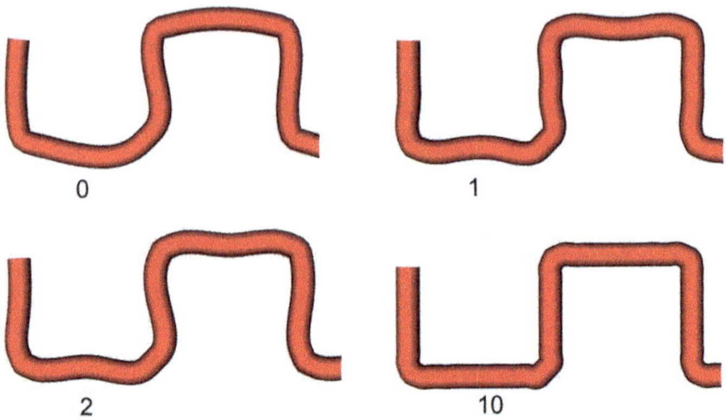

Fig. 8.5 Comparison of different interpolations. The number shows how many additional points are added between the key points, applying linear interpolation in order to get the straight parts less curved

The model can be extended to 12 points, if an additional point on the middle of each arc is placed (Fig. 8.3). The connecting line between the two stitched places remains still as an arc because of the interpolation. This should not be a problem, because the real form is also more similar to an arc than to a line, because of the compressibility of the fabrics. If a close to straight line is not wished, because it causes intersections with the textile in the initial configuration, additional points can be places on this line using linear interpolation.

The result of adding 1, 2 and 10 points between the keypoints using linear interpolation is visualized in Fig. 8.5. At 10 points between the key points the linear part is perfect, but the yarn starts to have additional curves around the arcs with the bend places. The optimal way of modelling such stitches would be to use analytic arcs and straight lines for smoothing the yarn path, but this would only be effective for small models and certain purpose. The implementation of smoothing models with curves and lines in generalized CAD software would increase the complexity of the data significantly, but is possible if the result requires it.

8.4 Lock Stitch—Class 301

The stitch type 301 is the most popular one. It consists of two yarns, which interlace somewhere in the middle or on the side of the sewed assembly (Fig. 8.6b). A set of keypoints, which can be used for its modelling, are presented in Fig. 8.6a. Here the points A, B, C, D, are the points at the beginning of the opening and the points in the interlacement area are numbered from 1 to 5 in Fig. 8.7. An important additional parameter is the vertical position of the interlacing point h_c. Optional are parameter h_1, which defines the height of the area of the loop and r_1. which allows the definition

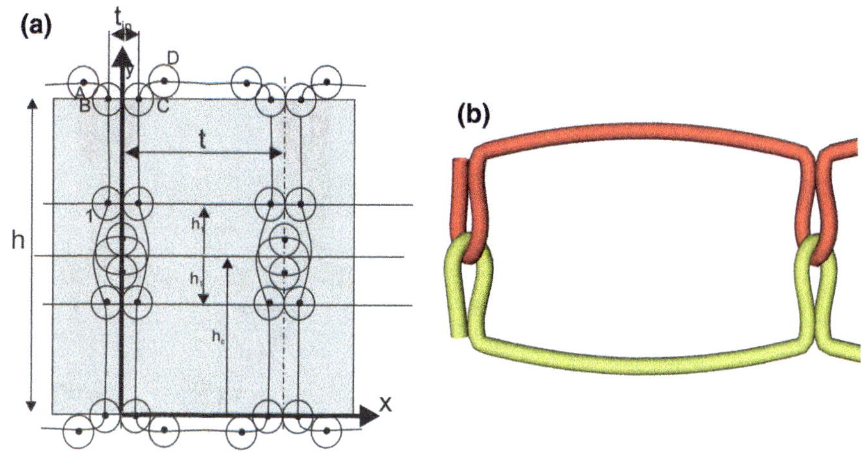

Fig. 8.6 Stitch 301 basic key points and 3D model

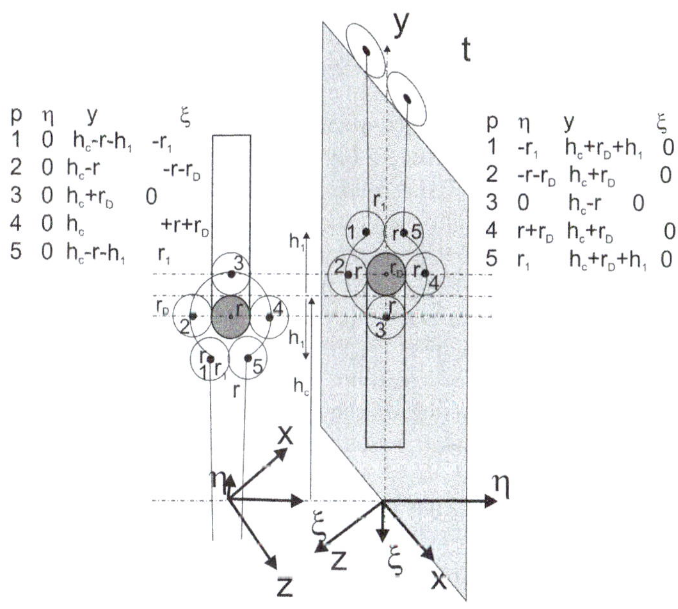

Fig. 8.7 Stitch 301—interlacing area and yarn coordinates there

of a larger distance between the yarns in this area. The radius of the bottom yarn is in this case r_D.

Both sewing yarns are crossing each other in the interlacing area. Crossing means that the planes where their axes are placed are almost perpendicular. These planes are rotated from the seam plane to some angle α. The angle α is also a parameter in the model. The description of the coordinates of the points is most efficient in the

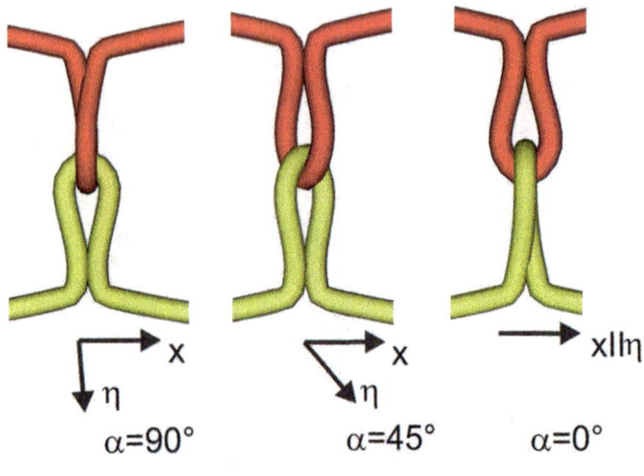

Fig. 8.8 Interlacement between the upper and bottom threads under different angle

local coordinates of the loop itself, where the y axis remains the same, the local x is rotated to some angle α (for instance $\alpha = +45°$) to the position η and the plane of the other yarn is then at ξ.

After the definition of the coordinates of the interlacing area in local coordinates (why two times coordinates?), these can be rotated using a simple matrix operation and are then added to the remaining unrotated points. The influence of the rotation of the interlacing area is visualized in Fig. 8.8 in which interlacements for three different angles are modelled. Using angle α allows the creation of a more correct geometry for the cases, in which one of the yarns is more stiff than the other—for instance in the case of $\alpha = 90°$ the bottom yarn can be the more stiff one or the yarn with higher tension, which does not rotate and the upper yarn is oriented around it.

The interlacement of the yarns can happen on one of the sides (up or down), if the difference of the yarn tension between the yarns is large. In this case, the change of the value of the parameter h_C could not be enough, because several points from the interlacing geometry are not needed any more. For example, points 1 and 5 can be removed, if the interlacement is close to the surface of the textiles (Fig. 8.9, middle case) or if the upper thread is straight, then of the upper yarn only the point 3 is required. The condition for this is based on the relation between the interlacing points between the sewing yarns (threads) and the fabric thickness. If the interlacing position is higher than the thickness:

$$h_C > h \tag{8.1}$$

than the points 1, 2, 4 and 5 from the coordinates of the upper thread have to be omitted. The situation is analogous, if the interlacing point moves under the fabrics:

$$h_C < 0 \tag{8.2}$$

Figure 8.10 demonstrates the results of the modelling of stitches in three interlacing positions.

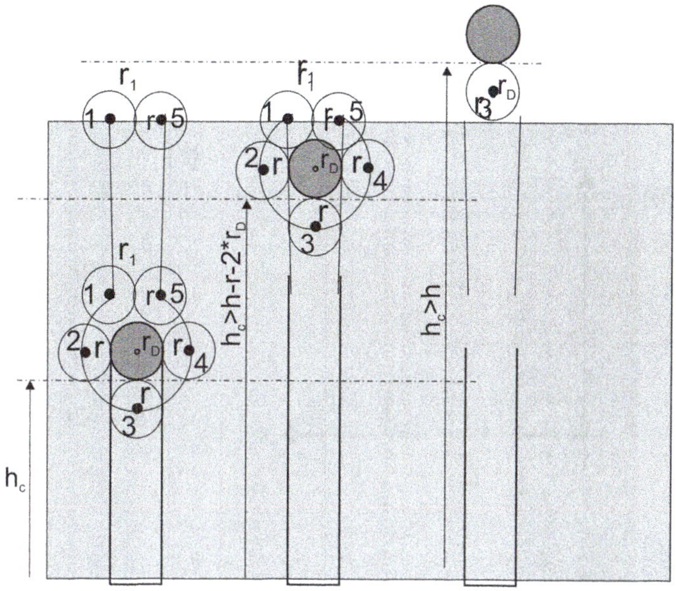

Fig. 8.9 Depending on the position of the interlacing points some of the keypoints become unnecessary and have to be removed. **a** Normal interlacement, **b** points 1 and 5, **c** points 1, 2, 4 and 5 should be removed

Fig. 8.10 Modelled stitch 301 with yarn interlacement on the upper side of the fabric, slightly under the middle level and under the sewed materials (these are not shown in order to keep the clarity of the figure)

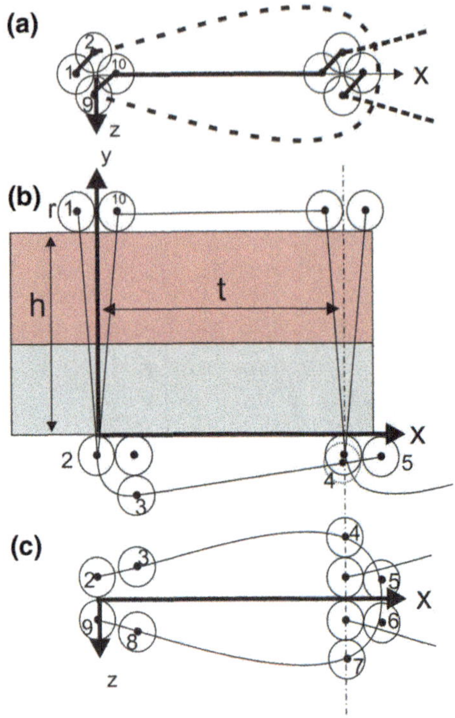

(d)

p	x	y	z
1	-r	t+r	0
2	0	-r	-r
3	2r	-3r	-r
4	t	-r	-3r
5	t+2r	-r	-r
6	t+2r	-r	+r
7	t	-r	+3r
8	2r	-3r	+r
9	0	-r	+r
10	+r	t+r	0

Fig. 8.11 Geometric model of the sewing stitch 100 **a** top view, **b** cross-sectional view, **c** bottom view, **d** local coordinates of the key points

8.5 Class 101

The lock stitch (class 101) consists of one yarn which interloops with itself. From the top side of the fabrics it appears straight, from the bottom part it builds loops in a similar configuration as the warp knitted loops. The variant of coordinates for the key points of such a stitch is presented in Fig. 8.11 with top, bottom and cross-sectional view. Figure 8.12 visualizes the modelled stitch one time with connecting lines between the key points (Fig. 8.12a) and one time using cubic spline spines with a few additional points in order to normalize the straight parts of the curve.

8.6 More Stitches

All other types of sewing stitches can be described and modelled a similar way. It is always important that the model includes enough parameters, which allow the modelling of different configurations of the yarns depending on their tension.

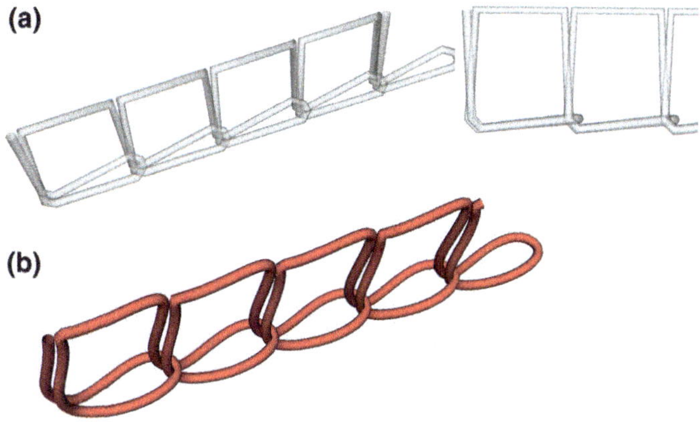

Fig. 8.12 3D model of the stitch 100. **a** Based on linear interpolation between the key points; **b** using spline interpolation and additional points for control the curve

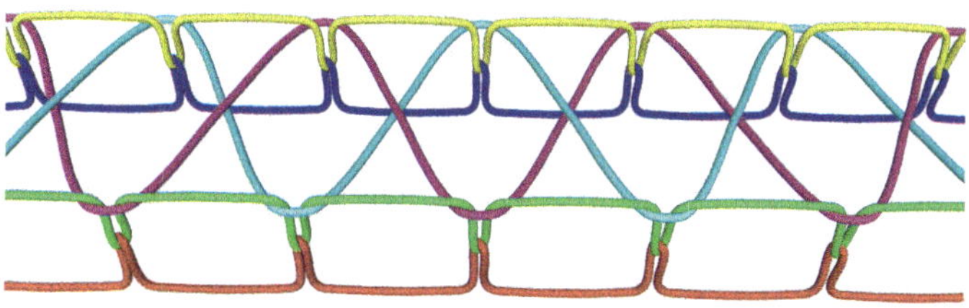

Fig. 8.13 3D model of the stitch based on two stitches 301 and two covering yarns [5]

Several examples of stitches of new computer controlled machines can be found in [1] (Fig. 8.13). These models are realized as python scripts and visualized with TexGen [12], while all previous images in the chapter are generated with the implementation of the algorithms in C++ as a software "TexMind Stich Generator".

8.7 Conclusions

This chapter demonstrates the principle of defining the coordinates of key points of some of the most commonly used sewing stitches. The coordinates are defined in a local coordinate system for each stitch, which allows its placement at any possible position and orientation in the space. The interlacing area between the yarns of the lockstitch (type 301) is placed in a separate local coordinate system, rotated by an angle relative to the local coordinate system of the stitch. This allows an additional adjustment of the interlacing area. All stitches have alternating areas with very small radii of curvature and straight areas. In order to ensure the correct interpolation of the paths between the topological key points, a few additional control points are included.

Due to the implementation of the described models in software it is now possible tto create the yarn geometry of stitches. These 3D geometries can be integrated into assemblies for the investigation of their properties using different computational models, or as well for the visualization during the teaching process.

References

1. Brinkert, N.: Experimentelle Evaluation der Funktionalität einer neuartigen Zierstich-Nähmaschine sowie 3D Modellierung der Stichvariationen mit der Software TexGen. Hochschule Niederrhein, Fachbereich Textil- und Bekleidungstechnik, Mönchengladbach (2016)
2. Floeck, M., Stadtfeld, H.C., Mitschang, P., Bickerton, S.: Impact of stitching processes on the compaction behavior of glass fiber reinforcements. J. Ind. Text. **36**(2), 151–165 (2016). https://doi.org/10.1177/1528083706068667
3. Koziol, M.: Effect of thread tension on mechanical performance of stitched glass fibre-reinforced polymer laminates—experimental study. J. Compos. Mater. **47**(16), 1919–1930 (2012). https://doi.org/10.1177/0021998312452179
4. Koziol, M.: Experimental study on the effect of stitch arrangement on mechanical performance of gfrp laminates manufactured on a basis of stitched preforms. J. Compos. Mater. **46**(9), 1067–1078 (2012). https://doi.org/10.1177/0021998311414947
5. Kyosev, Y., Brinkert, N., Zöll, K.: Strategy for simulating dry preforms for composites connected with sewing stitches. In: Lomov, S., Gorbatikh, L., Swolfs, Y. (eds.) COMPTest 2017 Leuven—8th International Conference on Composites Testing and Model Identification—Presentations. KU LEuven, Leuven (2017)
6. Ogale, A., Mitschang, P.: Compaction behavior of assembled fiber reinforced preforms. J. Ind. Text. **37**(1), 15–29 (2016). https://doi.org/10.1177/1528083707078195
7. Ogale, A., Mitschang, P.: Tailoring of textile preforms for fibre-reinforced polymer composites. J. Ind. Text. **34**(2), 77–96 (2016). https://doi.org/10.1177/1528083704046949
8. Rödel, H.: Kettenstiche. In: Gries, T., Klopp, K. (eds.) Füge- und Oberflächentechnologien für Textilien, VDI-Buch, pp. 14–18. Springer, Berlin (2007)
9. Sankaran, V., Younes, A., Engler, T., Cherif, C.: A novel processing solution for the production of spatial three-dimensional stitch-bonded fabrics. Text. Res. J. **82**(15), 1531–1544 (2012). https://doi.org/10.1177/0040517512452945
10. Singh, H., Mukhopadhyay, A., Chatterjee, A.: Influence of bias angle of stitching on tensile characteristics of lapped seam parachute canopy fabric—part i: mathematical modelling for determining test specimen size. J. Ind. Text. **46**(1), 292–319 (2015). https://doi.org/10.1177/1528083716631331
11. Tan, K.T., Watanabe, N., Iwahori, Y.: Impact damage resistance, response, and mechanisms of laminated composites reinforced by through-thickness stitching. Int. J. Damage Mech. **21**(1), 51–80 (2010). https://doi.org/10.1177/1056789510397070
12. TexGen (2007). http://texgen.sourceforge.net/index.php/Main_Page

Part V
Multiscale Modelling—Assemblies, Filaments and Their Software Implementation

Chapter 9
Extending to Filament Level and Interpolation Issues

9.1 Introduction

The fibers and yarns are one dimensional products—they are characterized by the coordinates of their axis and their cross section. Only drawing their axes as lines in the 3D space does not produce simulations of the product that are realistic enough, because the single lines have no material related porperties—they are not able to reflect light. This is the reason why the fibers and yarns are visualized and rendered as 3D surface. For several textile applications the already simplest ways of creating tubular surfaces along some paths show sufficient results. This task is trivial and is described in several books for computational geometry as well as in some textile related papers like [3–5, 14, 18, 20]) etc. Actually, the standard approach is not suitable for all types of textile products. This chapter presents the principles of the extending the yarn volume with filaments and some issues which cause troubles in this task.

9.2 Yarn Cross Section Form

Usually it was assumed that the yarn cross section is constant and circular. These assumptions are only close to reality for structures based on monofilaments or wires, but not for structures based on multifilaments or staple yarns. In this case the cross section of the yarn depends on several fiber and yarn parameters (twist, friction, fiber form—textured or not, etc.) as well as on the curvature of the yarn and the contact with the neighboring yarns. By bending the single filaments they are changing their position in the space to the most effective energy state and therefore, the cross section changes (Fig. 9.1a), [13]. As demonstrated in the structure in Fig. 9.1, a multifilament yarn can have a very compact cross section if compressed from all sides by other yarns (Fig. 9.1c). The cross section can deform to a flat shape at the bending points at which no other forces from the sides are applied (Fig. 9.1b). The situation is similar

Y. Kyosev, *Topology-Based Modeling of Textile Structures and Their Joint Assemblies*, https://doi.org/10.1007/978-3-030-02541-0_9

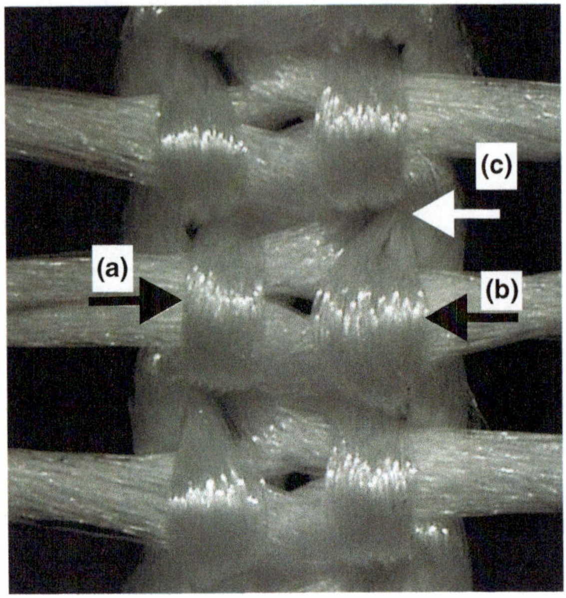

Fig. 9.1 Warp knitted fabrics made from multifilament yarns. The yarn cross section has a different form and size in different places. Reprinted from X. Chen (Ed.) "Modelling and predicting textile behaviour", Y. Kyosev, W. Renkens, "Modelling and visualization of knitted fabrics", pp. 225–262, Copyright (2010) with permission from Elsevier [13]

for dense woven structures, Fig. 9.2 shows multifilament woven belt from the top view and Fig. 9.3 shows its cross section.

These changes of the yarn cross section are part of the micro-scale modelling and usually require separate models and algorithms for a successful and efficient treatment. Examples of how the cross sections of multifilament yarns can be simulated with computational mechanic methods can be found in [21, 22], could be used in the future to refine the meso-scale model at several local places. Another approach—to

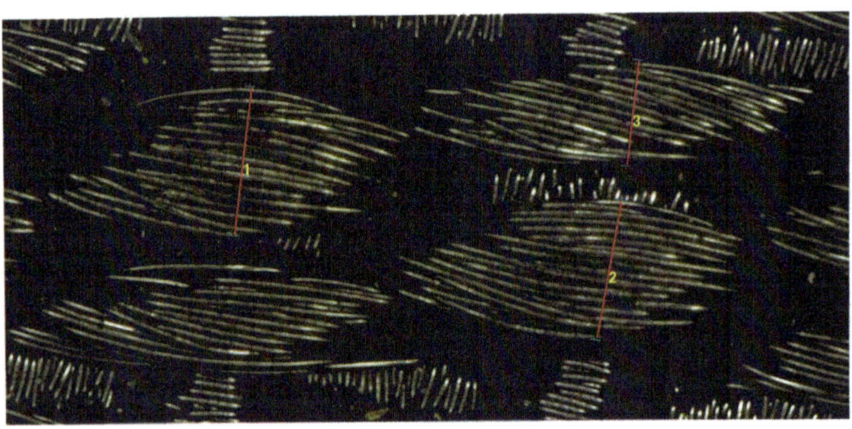

Fig. 9.2 Top view of woven fabric [8]

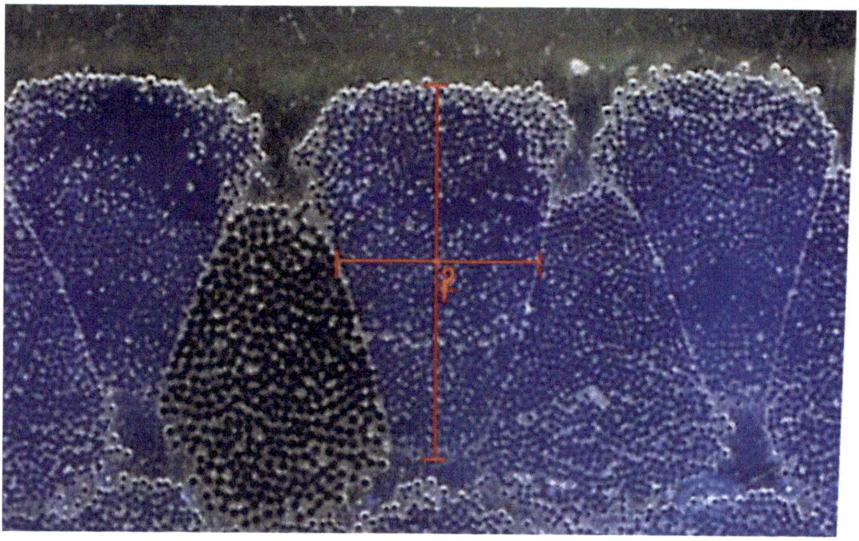

Fig. 9.3 Cross section of dense woven belt. From the top view only half of the yarns are visible [8]

model all filaments as single yarns—requires more intensive computations and is presented in [2, 19] etc. Current research papers in this area show examples with multifilament yarns with less then 20 filaments. This is noticeable less than in reality, but already allow the understanding of the interactions between the filaments. All these works and images demonstrate, that for proper yarn modelling the yarn cross section deformation have to be described correct. One method for this is the using of continuum based models for the yarns, as in [1] but in this case suitable "material model" for the yarns in lateral direction has to be defined. The yarns are actually no continuum, they are structures, consisting of multiple filaments. In this meaning, the more natural approach is to include this level—of the filaments in the models. For this the computed yarn volumes have to be filled by filaments (or fibers), as described in the next section.

9.3 Cross Section Definition for Multifilament Modelling

The 3D geometry, created for single yarns with circular cross sections can be used as a basis for the generation of single filaments within each of these yarns. The point of the curve of one loop is described as a curve, based on several points P (Fig. 9.4)

Referring to the rules of computational geometry it is well known, that tangent vector can be calculated as following

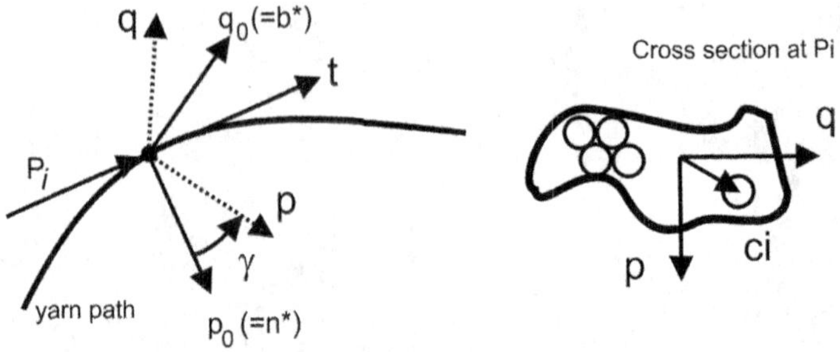

Fig. 9.4 Point of the 3D yarn axis and its characteristic vectors, vectors p and q determine the local coordinates of the cross section of the yarn at each point [12]

$$t\,(s) = \frac{p(s)'}{|p(s)'|} \tag{9.1}$$

And the binomial and normal vector to the curve at this point:

$$b\,(s) = \frac{p(s)' \times p(s)''}{|p(s)' \times p(s)''|} \tag{9.2}$$

$$n\,(s) = b(s) \times t(s) \tag{9.3}$$

As the yarn cross section can be rotated around the yarn axis, for instance due to twist of the yarn, additionally to the normal and binormal vector the vectors p and q are used, which define the local coordinate system of each cross section. These vectors are rotated by the angle γ, which is determined by the yarn twist, applying the rotation matrix R:

$$p\,(s) = R(s) \bullet t(s) \tag{9.4}$$

If the yarn cross section is defined as an arbitrary set of the center points of each filament c_i in a local coordinate system based on the vectors p and q, the positions of these centers can be determined after applying the rotation:

$$c_{i*}\,(s) = R(s) \bullet c_i(s) \tag{9.5}$$

Figure 9.5a presents loops of a yarn, created from a geometrical model and modelled with a constant cross section. Figure 9.5b presents the same loop filled with 15 filaments based on this method and Fig. 9.5c shows the cross-sectional view of the filaments. As it can be seen in Fig. 9.5b, there are some irregularities in the filament orientation which will be discussed in the next section.

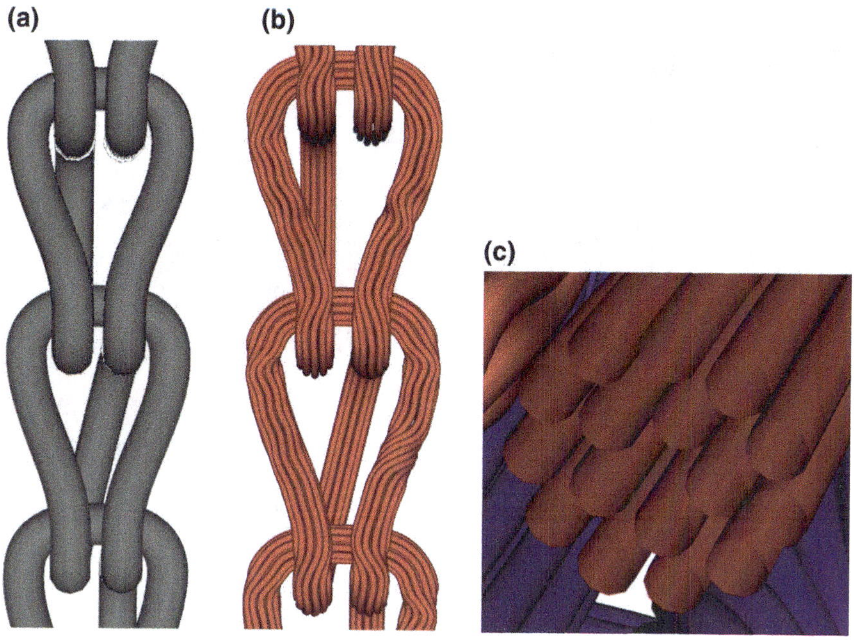

Fig. 9.5 Modelled warp knitted loop (**a**), filled with filaments (**b**), view of the cross section (**c**) [10]

9.4 Natural Curvature and Artificial Yarn Twist

Each curve in the 3D space has a natural curvature and torsion. The torsion of the curve shows the change of the binormal rotation around the tangent and is thus not connected with the twist of the yarn, which is using the same 3D curve as its axis. The twist is an additionally parameter, independent on the curve. At some places the orientation of the normal and binormal vector changes and this causes artificial twist of the filaments, which is caused only by the form of the curve. Such defects are visible in Figs. 9.5b and 9.6b and appears more drastically in areas where the radius of the curvature is too large and becomes indefinite. At such points the tangent vector can change its sign. This cause a complete revolution of the cross sections between two neighbouring axis points, as visibel in Fig. 9.6a.

This defect can be corrected by checking the sign of the tangent vector and in case this changes (Eq. 9.6), a sign correction is applied and the resulting normal and binormal vectors are calculated again after this correction.

$$if\ t(s_i) \bullet t(s_{i-1}) < 0\ then\ t(s_i) = -t(s_i) \tag{9.6}$$

The orientations of the normal and the binormal vectors have to be re-calculated after a change in the tangent vector. When the curve has zero curvature and is a

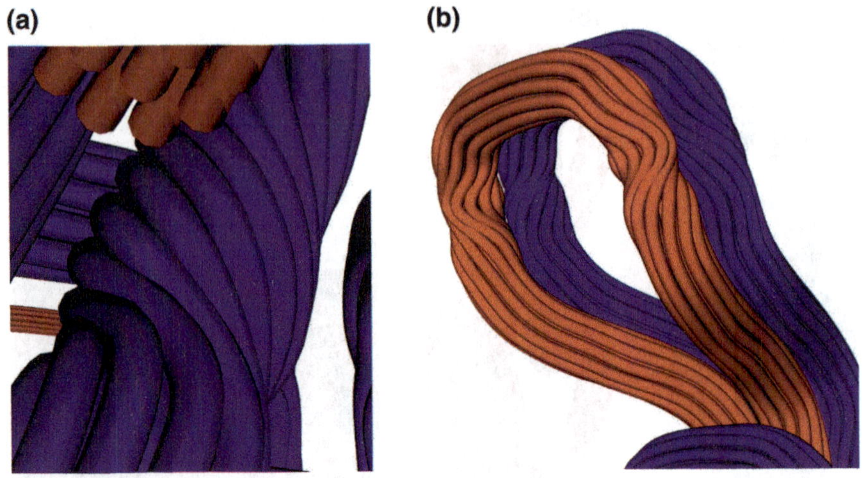

Fig. 9.6 Defects of the model, based on the torsion of the 3D curve, [10]

straight line, the vectors for the next point can be assumed to be equal to these of the previous one point:

$$\mathbf{n_i} = \mathbf{n_{i-1}} \tag{9.7}$$

$$\mathbf{b_i} = \mathbf{b_{i-1}} \tag{9.8}$$

9.5 Interpolation Issues

The main principle for the creation of the yarn topology is based on the definition of several points on the yarn axis, which coordinates can be calculated parametrically. One part of the models does not only use the points, but defines the curves and lines between these points explicitly. Figure 9.1a presents a well-known model for woven fabrics produced from yarns with circular cross sections, in which each yarn axis can be presented by a connected circle arcs and straight lines (Fig. 9.7a). An extended version of such models is successfully implemented in the commercially and academically used software Wisetex [15–17] for woven structures. The large number of models for weft knitted structures is reported by Kurbak and Ekmen [6], Kurbak and Soydan [7] and geometries rendered within 3DS studio are demonstrated. If these models can be applied for general cases within textile CAD software, depends on the sensibility of the models and on the input data, which should be tested in the future. Less accurate but more flexible for the implementation in computer applications are the models based on the sequence of the key-points only, in which the connecting curve is computed at a next step. A drawing of the development of such models for woven structures is presented in Fig. 9.7b.

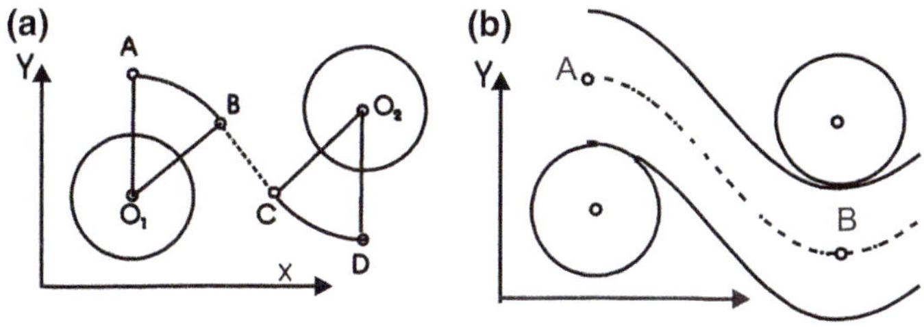

Fig. 9.7 Idealized geometrical model of the yarn path of a woven structure **a** based on circular arcs and lines **b** based on spline interpolation between key points. Reprinted from D. Veit (Ed.) "Simulation in Textile Technology", Y. Kyosev, "Simulation of wound packages, woven, braided and knitted structures", pp. 266–314, Copyright (2012) with permission from Elsevier [11]

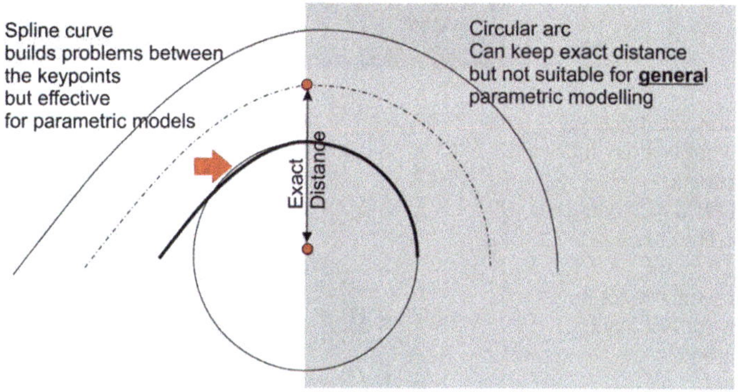

Fig. 9.8 The inerpolation with splines does not guarantee the path of the curve between the key points and can lead to intersections between the volumes. Using predefined functions (right hand side) solves this problem, but is not suitable for general modelling

The flexibility of these models is in the lower number of relations, which have to be programmed. This allow the extension of these models with additional structural elements to be done in more clear way. For woven structures, one and the same algorithm can be used for a complete class of patterns. For knitted structures the programming of loops, tucks, missed and transferred stitches can be done with less steps. Even though is has a lot of advantages, this level of simplicity has its price. During the interpolation the curve between the key points, the yarn volume can intersect the neighbour yarns at some positions, as visualized in Figs. 9.8 and 9.9. This error occurs, because the distance between the yarns is only defined at the places between the key-points and what happens between these key-points is not controlled during the generation of the interpolating curves.

The choice of a suitable modelling approach depends on the specific case and it needs to be decided by the modeller. For cases, in which more generality is required and in which several classes of structures have to be covered with one algorithm, the use of topological key point lists has definitive advantages. For models, created to investigate **one** type of structure (during Master or PhD thesis or investigation of one specific product) the analytical description of the curves can be preferable.

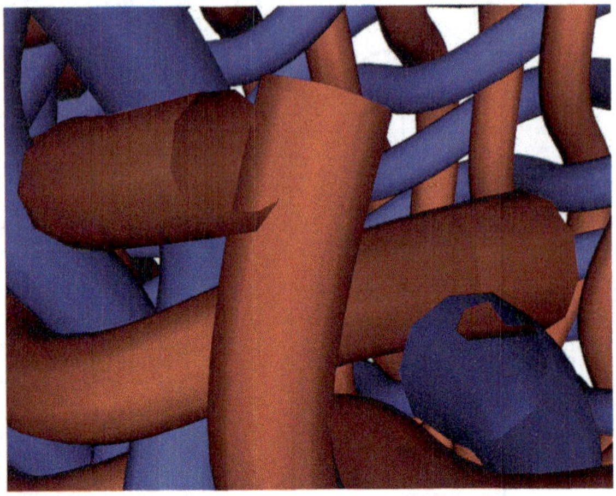

Fig. 9.9 Interpenetration between yarn volumes in the interpolated segments between the key points

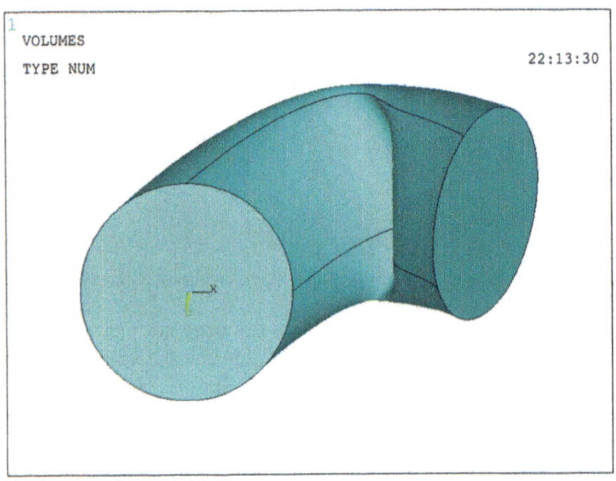

Fig. 9.10 Generated volume by sweeping circle along the yarn axis within a commercial FEM software (year 2009). The curvatures for the typical textile applications seems to not be integrated in the FEM and CAD software [9]

9.6 Too High Resolution at High Curvature Areas

In the textile structure the fibers and yarns build arcs with a very small radii compared to their length. As the radius of the curvature is often similar to the yarn diameter, most of the "normal" CAD systems have problems calculating the volume representation of such paths. Even when some results are produced, in most cases they are not applicable for textile applications (Fig. 9.10).

For this reason, the textile software uses its own rendering procedures or suitable libraries of computer graphics, for which this problem is recognized and solved. The reduction of the number of calculated cross sections so that they do not cross each other in such arcs, normally solves the problem [17].

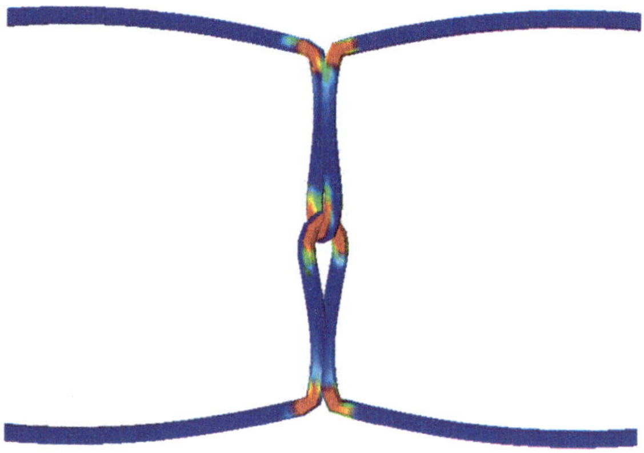

Fig. 9.11 Sewing stitches and warp knitted fabrics have areas with highly alternating curvature and areas with zero curvature (straight) lines. Here curvature of sewing stitch in colour

9.7 Highly Different Curvatures

Common problems during the computing of the surface of a single yarn or during computing the orientation and position of the filaments in a yarn are the structures, in which the curvature of the yarn axis changes rapidly. This is a common issue for knitted structures. The loops have places with a curvature radius equal to the yarn diameter and also floatings or underlapps which are straight lines or show zero curvature. A similar situation can be deteced for sewing stitches. At the interlacing point between two yarns these have a radius of curvature equal to the yarn diameter (very small) and between the stitches on the fabrics surface the yarns are straight lines (endless radius of curvature) (Fig. 9.11)

9.8 Interpolated Areas

If only the basic points used for topological representation are also used for smoothing the yarn axis using spline, the resulting configuration can become wrong (Fig. 9.12). For the visualization of the case presented in Fig. 9.12 the underlapping is represented as just a straight line. The red yarn (R) is between the blue yarn (B) and the loops, and this is the correct place of it. If a spline is used for smoothing (Fig. 9.12b) of this pattern, the distance between the two key points of the red underlapping is greater than the distance for the red yarn and the curve becomes larger. The red yarn moves outside the fabric, which is wrong. For the solving this problem additional points along the underlapping have to be used. They do not improve the topology, but help to produce correct results in the interpolated segments

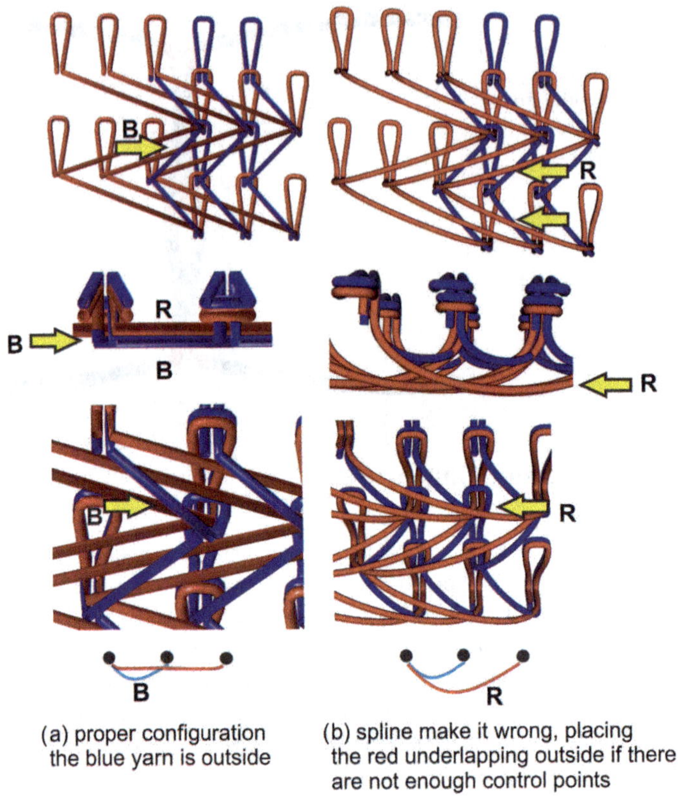

Fig. 9.12 Problem caused by interpolation with not enough control points. **a** Proper orientation of the straight yarn pieces (underlapping)—B is outside, **b** wrong representation after applying spline (or any other function with equal first derivative from the both side of the control points)—R is outside

9.9 Solution Strategies

There are different possibilities to obtain the correct path of the yarn axis in one structure. This can be done by selecting suitable functions for the different regions, controlling the values of the tangent vectors or adding additional control points.

The most stable method uses different kinds of functions specially prepared for the different segments, as already demonstrated with the circular arc in Fig. 9.8. For this method, each part of the yarn axis has to be approximated with different curve. The example in Fig. 9.13a needs three straight lines (polynomial of degree 1), one polynomial of a degree of 2 or 3, or trigonometric functions need to be used to represent the arcs. This way can be very effective for woven structures, because their yarn axes are mostly plain curves (initially) [17]. This approach is applied in some research papers for weft knitted structures [6, 7], but is normally not effective for generalized CAD software in which the yarns are spacer curves. The list of the

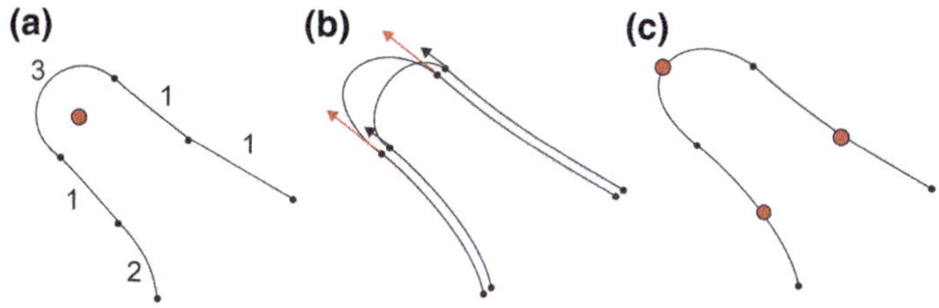

Fig. 9.13 Problem caused by interpolation with too little control points. **a** Proper orientation of the straight yarn pieces (underlapping)—B is outside, **b** wrong representation after applying spline (or any other function with equal first derivative from the both side of the control points)—R is outside

different types of curves will cause problems in the speed and the calculation of the volume, because for each of these curves the derivatives have to be defined, available, and determined in the area. Another solution is the direct definition of the tangent, normal and binormal vectors for each point as done for several 3D structures in Wisetex [17].

The application of the cubic splines allows unified calculation of all derivatives and segments. This allows the usage of one visualization library for the yarns and fibers, which is independent from the structure definition. The path of the cubic splines between the key-points can be controlled using the values of the tangents at these points (Fig. 9.13b). The identification of the proper value of the tangent can be done using optimization procedure, but it is not commonly done by the textile engineers. The simplest and most efficient method in this context is the usage of additional points on the paths, in all areas of the segments, which the yarn should cross (Fig. 9.13c). These additional control points can guarantee that the interpolated segments of the curve pass through the proper places and the interlacement of the yarns remains correct for the investigated structure.

9.10 Examples

The following three images present modelled structures at the filament level. Each structure is first created at yarn level. After that, the yarn axis is used as the leading curve for the determination of the coordinates of the single filaments. If filaments only have to be placed on the surface, the relations are simple. Figure 9.14 presents such three braids.

In some cases, as electric cables or other complex structures, the yarns are made from more twisted filaments. For such structures it is possible to define the local coordinates of the filaments in circular layers (Fig. 9.15). Each layer can have different numbers and types of filaments.

In the context of composite structures the flat cross sections (Fig. 9.16) are most common. In such cases the cross section can be defined as several parallel layers of filaments, each with separated vertical coordinates, numbers of filaments and distance between these.

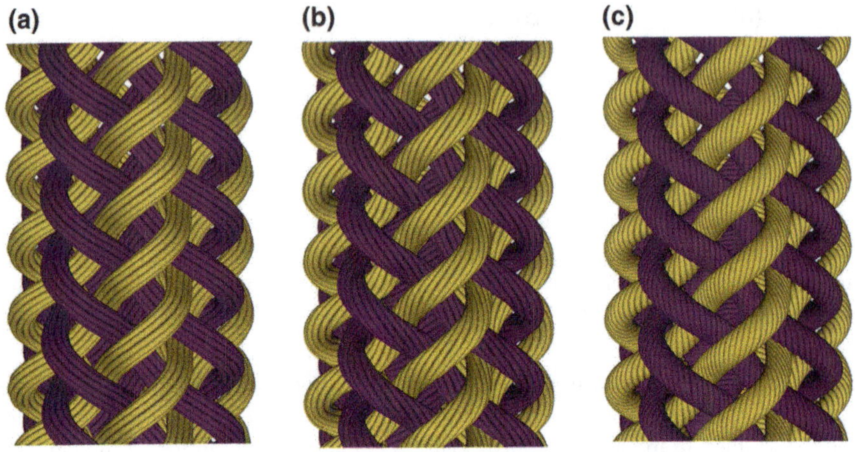

Fig. 9.14 Braids, modelled with filaments in the yarns. **a** Without twist, **b** and **c** with different twist

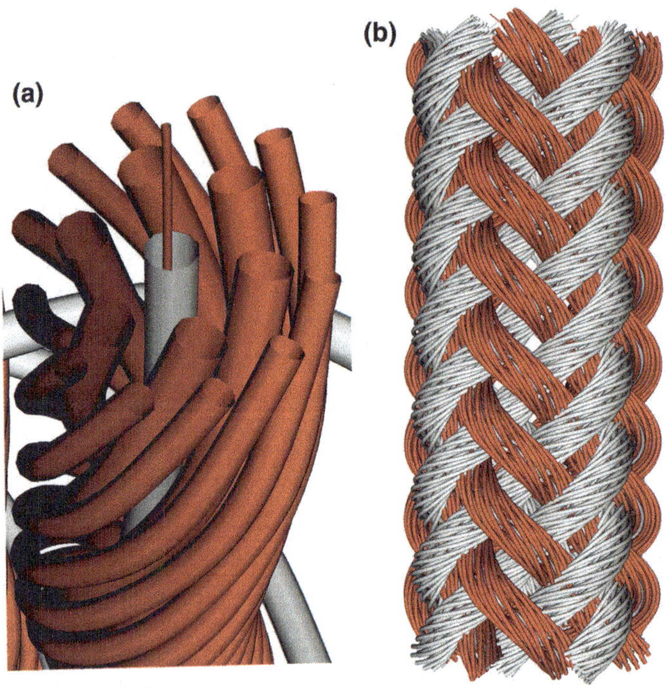

Fig. 9.15 Example of multilayered circular cross section with filaments. Each layer has a nominal radius, a filament radius and a number of filaments

Fig. 9.16 Example of multilayered flat cross section. Each layer has its own vertical coordinate, number of filaments, filament diameter and distance between the filaments

9.11 Conclusions

The coordinates of the solid cross section boundary and the coordinates of the single filaments are calculated in an identical way and show the same problems in the modelling. The curvature of the yarn axes in the 3D space is ranging between large values—segments with very small radius of curvature are alternating with parallel segments. This is causing problems with the definition of the normal and binormal vector at the points where some of the derivatives of the equation of the curve are undefined. These places are causing artificial twisting effects and become visible in both, the visualization of single yarns as a tube and in the representation of the yarn cross section as a set of filaments. Another issue is that the interpolation of the yarn axis with polynomials does not ensure the configuration of the curve between the key points. In some configurations, the interpolated curve leads to a wrong topology. This chapter points out several of such problems and demonstrates some possible solutions for them.

References

1. Boisse, P., Gasser, A., Hagege, B., Billoet, J.L.: Analysis of the mechanical behavior of woven fibrous material using virtual tests at the unit cell level. J. Mater. Sci. **40**(22), 5955–5962 (2005). https://doi.org/10.1007/s10853-005-5069-7
2. Durville, D.: Simulation of the mechanical behaviour of woven fabrics at the scale of fibers. Int. J. Mater. Form. **3**(S2), 1241–1251 (2010). https://doi.org/10.1007/s12289-009-0674-7
3. Goktepe, O., Harlock, S.: Three-dimensional computer modeling of warp knitted structures. Text. Res. J. **72**, 266–272 (2002)

4. Hardt, K.: Cyros—what it is and what it does. die bedeutung und arbeitsweise von cyros. In: Proceedings of the Cotton Incorporated 9th Annual Engineered Fiber Selection System Conference Research Triangle Park, USA, May 20–22, pp. 185–202 (1996)
5. Hardt, K.: Three-dimensional representation of filament structure as an aid in the development of technical fabrics. In: 7. International Techtextil Symposium, Neue Verbundtextilien und Composites, Produktions- und Verarbeitungstechnik, Messe Frankfurt GmbH, Frankfurt/M, D, 19–21. Jun, 1995, Band 3.3 (1995) Seite 1-3, Paper-Nr. 3.36 (1995)
6. Kurbak, A., Ekmen, O.: Basic studies for modeling complex weft knitted fabric structures part i: a geometrical model for widthwise curlings of plain knitted fabrics. Text. Res. J. **78**(3), 198–208 (2008). https://doi.org/10.1177/0040517507082352
7. Kurbak, A., Soydan, A.S.: Basic studies for modeling complex weft knitted fabric structures part iii: A geometrical model for 1 × 1 purl fabrics. Text. Res. J. **78**(5), 377–381 (2008). https://doi.org/10.1177/0040517507082465
8. Kyosev, Y., Lomov, S., Küster, K.: The numerical prediction of the tensile behaviour of multilayer woven tapes made by multifilament yarns. In: Kyosev, Y. (ed.) Recent Developments in Braiding and Narrow Weaving, pp. 69–80. Springer International Publishing, Cham (2016). https://doi.org/10.1007/978-3-319-29932-7_7
9. Kyosev, Y., Renkens, W.: Geometry preprocessing of textile reinforced composites for ANSYS. In: ANSYS Conference & 28th CADFEM Users' Meeting 2010 (2010)
10. Kyosev, Y.: Modelling of warp knitted structures at filament level. In: Avadanei, M. (ed.) 16th Romanian Textiles and Leather Conference CORTEP 2016, pp. 114–119 (2016)
11. Kyosev, Y.K.: Simulation of wound packages, woven, braided and knitted structures. In: Veit, D. (ed.) Simulation in Textile Technology, pp. 266–309. Woodhead Publishing Limited (2012). https://doi.org/10.1533/9780857097088.266
12. Kyosev, Y.K.: Generalized geometric modelling of tubular and flat braided structures with arbitrary floating length and multiple filaments. Text. Res. J. **86**(12), 1270–1279 (2015)
13. Kyosev, Y., Renkens, W.: Modelling and visualization of knitted fabrics. In: Chen, X. (ed.) Modelling and Predicting Textile Behaviour, pp. 225–262. Woodhead Publishing and In association with the Textile Institute and CRC Press, Cambridge, Boca Raton, FL (2010)
14. Liao, T., Adanur, S.: 3D structural simulation of tubular braided fabrics for net-shape composites. Text. Res. J. **70**(4), 297–303 (2000). http://trj.sagepub.com/cgi/content/abstract/70/4/297
15. Lomov, S., et al.: Wisetex (2011). http://www.mtm.kuleuven.be/Onderzoek/Composites/Research/meso-macro/textile_composites_map/textile_modelling/textile_modelling_fe
16. Lomov, S.V., Willems, A., Vandepitte D., Verpoest, I.: Simulations of shear and tension of glass/pp woven fabrics. In: Juster, N., Rosochowski, A. (eds.) The 9th International Conference on Material Forming ESAFORM, pp. 783–786 (2006)
17. Lomov, S.V.,: Integrated Textile Preprocessor WiseTex, Ver.3.2—Computational Models, Methods and Algorithms. KU LEuven, Leuven (2013)
18. Lomov, S., Gusakov, A., Huysmans, G., Prodromou, A., Verpoest, I.: Textile geometry preprocessor for meso-mechanical models of woven composites. Compos. Sci. Technol. **60**(11), 2083–2095 (2000)
19. Mahadik, Y., Hallett, S.: Finite element modelling of tow geometry in 3D woven fabrics. In: Advani, S.G., Gillespie, J.W. (eds.) Recent Advances in Textile Composites. DEStech Publications, Lancaster, PA (2008)
20. Pastore, C.M., Birger, A., Clyburn, E.: Geometrical modelling of textile reinforcements. In: Poe, C.C., Harris, C.E. (eds.) Mechanics of Textile Composites Conference, pp. 597–623. VA, Hampton (1995)
21. Samadi, R., Robitaille, F.: Particulate Methods for the Mechanics of Dry Textiles: Compaction of Yarn Assemblies. In: Advani, S.G., Gillespie, J.W. (eds.) Recent advances in textile composites. DEStech Publications, Lancaster, PA (2008)
22. Samadi, R., Robitaille, F.: Particle-based modeling of the compaction of fiber yarns and woven textiles. Text. Res. J. **84**(11), 1159–1173 (2014). https://doi.org/10.1177/0040517512470200

Chapter 10
Assembly Level—From Textile Structures to Textile Assemblies

10.1 Introduction

The previous chapters describes the rules for creating topological models of single textile structures and sewing stitches. The modelling of the structures is important as it allows efficient investigation of their properties and improvement of their design. Once one or several structures are optimized for some use, they are normally joint together into a textile assembly. This can be a cloth, profile for composite or concrete reinforcement, replacement of human part, textile sensor or another part of a complex object. This chapter describes the methods which allow to built a 3D geometry of the assembly, keeping the topological details at the yarn (and fiber) level. These methods are well known and used in the computer graphics, but in the following these are summarized with the relation to textile objects.

10.2 Structural Cells

All textile structures can be represented at the topological level as a set of structural cells, in which the relations between the yarn coordinates are defined. These structural cells can be **unit**-cells, **repeats**—in the context of patterning-, or larger group of cells. For the representation at the assembly level the smallest cell has to represent the important relations between all yarns which are crossing it. Figure 10.1 represents such cells for the different types of structural elements.

Each single element consists of one or multiple yarns. The coordinates of the points of each yarn are described in a discrete form as a vector of point coordinates in the **local** coordinate system in a form

Y. Kyosev, *Topology-Based Modeling of Textile Structures and Their Joint Assemblies*, https://doi.org/10.1007/978-3-030-02541-0_10

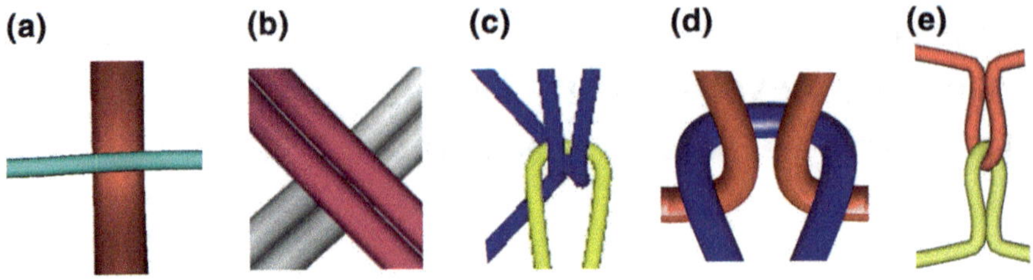

Fig. 10.1 Basic cells of the common structural elements **a** woven, **b** braided, **c** weft knitted, **d** warp knitted, **e** stitch

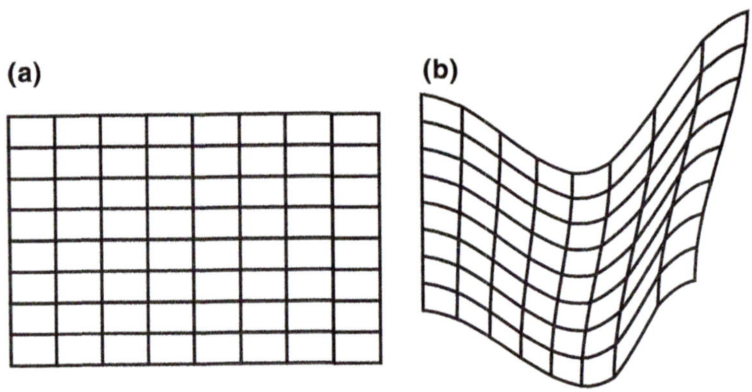

Fig. 10.2 Mesh of the structural cells at mezo level in initial **a** and deformed **b** state

$$C = \begin{bmatrix} x_1 & y_1 & z_1 & u_1 \\ x_2 & y_2 & z_2 & u_2 \\ & \cdots & & \\ x_{m-1} & y_{m-1} & z_{m-1} & u_{m-1} \\ x_m & y_m & z_m & u_m \end{bmatrix} \tag{10.1}$$

where x_i, y_i, z_i, u_i are the homogeneous coordinates of the i-th point of the yarn segment, consisting of m points. The homogenous coordinates of a point of a space with k dimensions build a vector of length $k+1$ and they allow projective transformations to be represented easily by matrix operations, where affine transformations can be simply combined by multiplying successive matrices. The Cartesian coordinates of a point with homogeneous coordinates (x, y, z, u) are (x/u, y/u, z/u). This simplifies and speeds up the calculations in the real-time graphics systems.

The data for each structural cell is stored in an array. In the object oriented programming languages (and the modern numerical computing environments like Matlab) it is not a problem to have two- or more-dimensional array of elements (cells), which contains variable numbers of vectors as point coordinates.

If the plane of the fabric is rotated, translated or deformed in some way, the global coordinates of the cells change (Fig. 10.2).

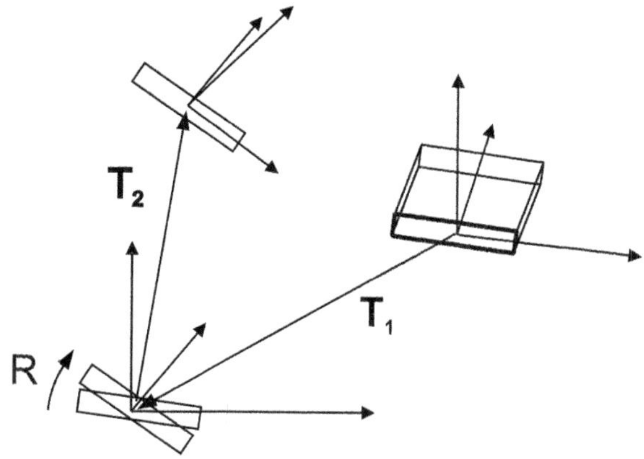

Fig. 10.3 Sequence for rotation (and translation) of a cell, if its components are stored in global coordinates. Renkens, W. and Kyosev, Y., Geometry Modelling of Warp Knitted Fabrics With 3d Form, Textile Research Journal, 2011, Vol. 81 no.4, pages 437–443. Copyright (C) 2011 SAGE. Reprinted by permission of SAGE Publications, https://doi.org/10.1177/0040517510385171, [8]

Each cell has then to be moved, rotated or deformed in a suitable way, but these operations are simple matrix operations, which can be performed in the local cell coordinates for all elements of the cell. For instance, the rotation of the cells in the local coordinate system around or the x-axis on an angle α_x can be done by multiplication with the rotation matrix R :

$$R = \begin{bmatrix} 1 & 0 & 0 & 0 \\ 0 & \cos(\alpha_x) & \sin(\alpha_x) & 0 \\ 0 & -\sin(\alpha_x) & \cos(\alpha_x) & 0 \\ 0 & 0 & 0 & 1 \end{bmatrix} \tag{10.2}$$

The storage of the coordinates of each structural element in local coordinate system brigns a major advantage, as all rotations and translations of the elements are simple. This method allows the multiscale modelling of different 3D structures, developed the time before, for example with the weft knitting technique. The method of the production of such structures is described in [1] and some new results are reported in [3, 11].

If the the yarn coordinates are stored directly in the global coordinate system, then the rotation of the single elements is more complicated. In such case the cells first have to be moved to the origin of the coordinate system (translation $T1$), rotated and than moved back to its previous or new position using translation $T2$ (Fig. 10.3)

$$C_1 = T_2 \cdot (R \cdot (T_1 \cdot C)) \tag{10.3}$$

where the brackets have to signify that the sequence of the operations is important and beforehand T_1 has to be applied, followed by R and then T_2.

The translation matrices have the following structure:

$$T = \begin{bmatrix} 1 & 0 & 0 & Tx \\ 0 & 1 & 0 & Ty \\ 0 & 0 & 1 & Tz \\ 0 & 0 & 0 & 1 \end{bmatrix} \tag{10.4}$$

where *Tx*, *Ty* and *Tz* are components of the translation vectors.

This method is developed initially for warp knitted structures [8] can applied for the following cases:

- fabrics with constant cross section (such as extruded forms) among the product axis, which could be produced with:
  - constant thickness (spacer fabrics) (for application in textile reinforced concrete [9], or seats, rucksacks, shoes etc.)
  - stepwise constant thickness (spacer fabrics with different thickness, [4])
  - hollow structures, for medical applications or other hose type products
- fabrics with variable cross section in the XZ-plane
- variations of the above, but with additional local surface relief.

10.3 Application for 3D Structures

10.3.1 Spacer Fabrics with Stepwise Different Thickness

Fabrics with controlled thick and thin places (stepwise different thickness) can be produced directly on a warp knitting machine without sewing, as developed and patented by Helbig et al. [4]. The idea of the patent is that the fabric is knitted in sections with different pile lengths and the proper form is built after the knitting process. The length of the connecting pile yarns between the loops of the front and rear beds determine the thickness of this part after knitting.

There are two basic ways to simulate such structures:

- directly in its relaxed form (Fig. 10.6) or
- as whole process, starting from the production of the sample, including all additional transformations and modifications and finally the relaxation process.

The first aproach is quicker for single products, but not in general case. Thus, the second way is described in the following.

First, the fabric is simulated in the form in which it is knitted on the machine (Fig. 10.5). Then the transformation matrix is determined. Hereonly the translation with vector **T** is required. The vector **T** is determined according to Fig. 10.4, using the simple geometry

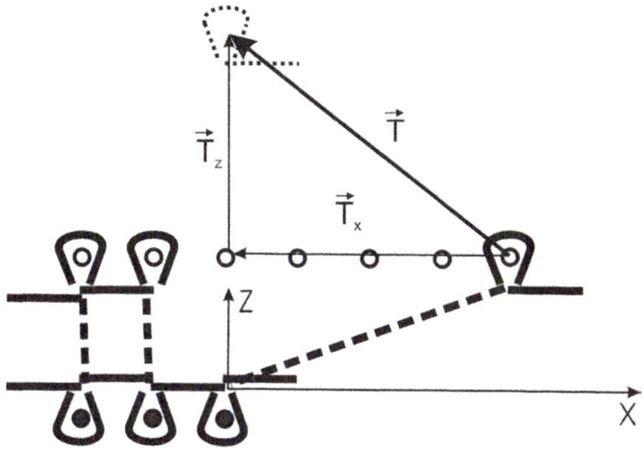

Fig. 10.4 Translation of loop or a cell to build structure with stepwise thickness. Renkens, W. and Kyosev, Y., Geometry Modelling of Warp Knitted Fabrics With 3d Form, Textile Research Journal, 2011, Vol. 81 no.4, pages 437–443. Copyright (C) 2011 SAGE. Reprinted by permission of SAGE Publications, https://doi.org/10.1177/0040517510385171, [8]

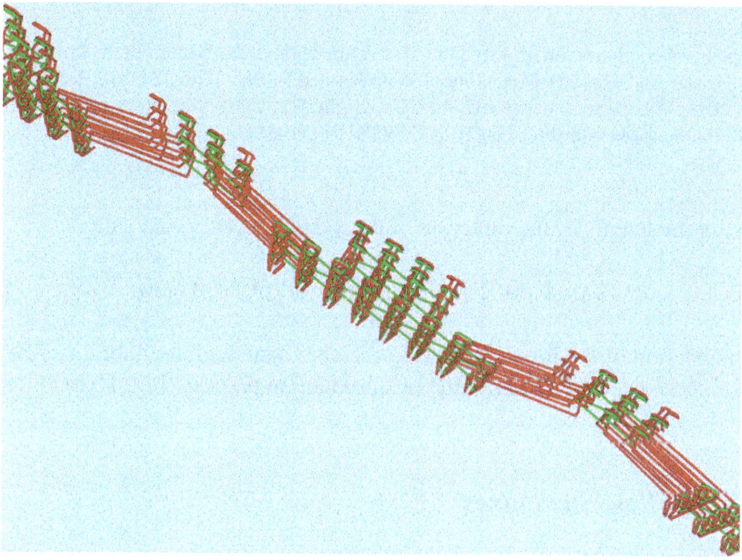

Fig. 10.5 Initial configuriation (on the machine) of a warp knitted structure for production of areas with different thickness. Renkens, W. and Kyosev, Y., Geometry Modelling of Warp Knitted Fabrics With 3d Form, Textile Research Journal, 2011, Vol. 81 no.4, pages 437–443. Copyright (C) 2011 SAGE. Reprinted by permission of SAGE Publications, https://doi.org/10.1177/0040517510385171 [8]

$$|\mathbf{T}| = \sqrt{A^2 + (n \cdot w)^2}, \tag{10.5}$$

where A is the distance between both needle beds, n—the number of needles, above which the loops are shifted to the right and w—the distance between the needles.

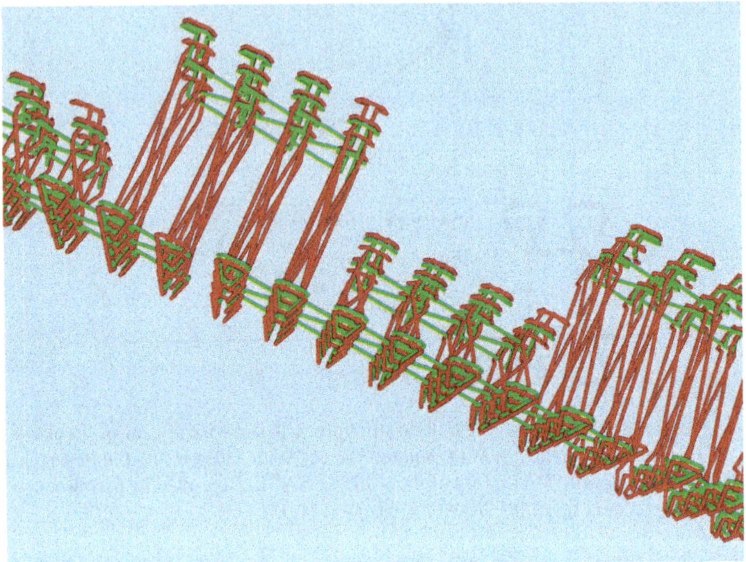

Fig. 10.6 Final configuration of warp knitted structure with areas of different thickness. Renkens, W. and Kyosev, Y., Geometry Modelling of Warp Knitted Fabrics With 3d Form, Textile Research Journal, 2011, Vol. 81 no.4, pages 437–443. Copyright (C) 2011 SAGE. Reprinted by permission of SAGE Publications, https://doi.org/10.1177/0040517510385171 [8]

Having the length T, the direction of the vector can be found using:

$$\mathbf{T} = \mathbf{T}_X + \mathbf{T}_Z = -n \cdot w \cdot \mathbf{e_x} + (|\mathbf{T}| - A) \cdot \mathbf{e_Z} \tag{10.6}$$

After translation of the loops, following the above method, the fabric can be represented in the form depicted after the postproduction process (Fig. 10.6).

10.3.2 Hollow Structures

Hollow structures are another type of three dimensional warp knitted fabrics, which are used in medicine as artificial veins, ropes, and in the food industry e.g. net coverings for salami. They can be produced

- on double bed warp knitting machines with three systems of yarns–one on the fabric part on the front bed, one for the part on the rear bed and one yarn system for the building of connections between both parts;
- flat knitting machines with two needle beds, where one yarn builds rows in the front and then on the back;
- on circular knitting machines.

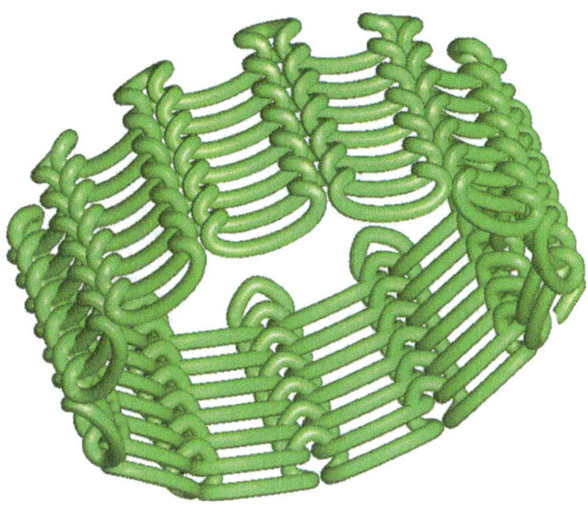

Fig. 10.7 Tubular weft knitted, single face fabric

All these structures are build based on topological structual cells which are placed around a cylindrical surface. Each segment has to be rotated to angle

$$\alpha_i = \alpha_0 + \frac{360^\circ}{N} \cdot i \tag{10.7}$$

where N is the total number of the cells in the periphery of the structure and $i = \{1, 2, ..., N\}$, which corresponds to the number of the needles . $\alpha_0$depends on which needle has been selected to be first and where its loop will be positioned in the space. An example of a structure modelled in this way is presented in Fig. 10.7. It is assumed, that the product radius and the loops main sizes are known prior to the simulation. If this is not the case, additional calculations for the estimation of the shrinkage of the loops and the effective rope radius have to be done, in order for the simulated structure to correspond to the real one.

10.4 Assembly Position and Orientation

For the modelling of the complete assemblies, the same rule as for the single cells applies. The mesh of each part has to be moved, rotated, deformed to the final position. After that the orientation and the position of each cell is used for the calculation of the global coordinates of the yarns based on the local cell coordinates, applying the matrix transformations. The matrix transformations are well known in the computer graphics and are described for instance in [2] or in almost any other book about computer graphics. A software realization for the adjustment of a stack of unit cells for composites is the LamTex package from the WiseTex suit [6, 7]. In Fig. 10.8 the rotation and transformation of a piece of warp knitted fabric is demonstrated.

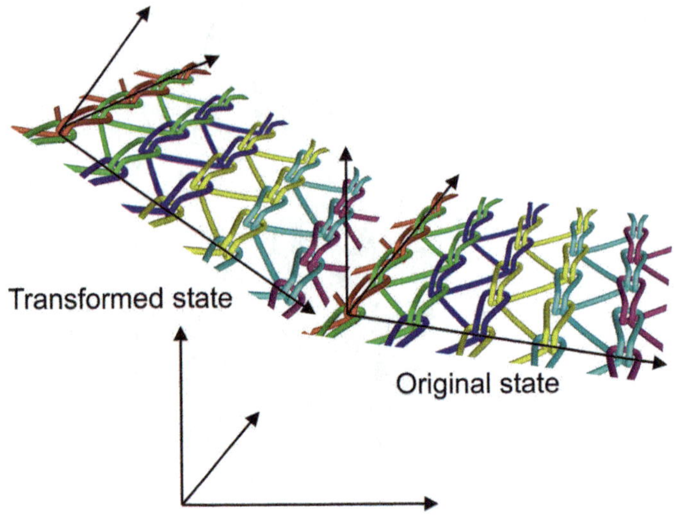

Fig. 10.8 Warp knitted fabrics in initial state (for instance after import) and after application of some transformation (rotation and translation). Renkens, W. and Kyosev, Y., Geometry Modelling of Warp Knitted Fabrics With 3d Form, Textile Research Journal, 2011, Vol. 81 no.4, pages 437–443. Copyright (C) 2011 SAGE. Reprinted by permission of SAGE Publications, https://doi.org/10.1177/0040517510385171 [8]

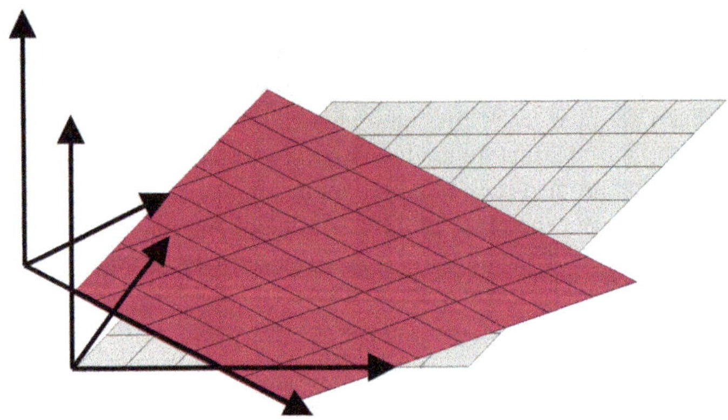

Fig. 10.9 Adjusted meshes of two fabrics prepared for stitching—additionally to the transformations for getting the proper orientation (like rotation and translation) also the cutting of some of the fabrics is applied in order to achieve some pattern

10.5 Cutting of Fabrics

When assemblies are produced the fabrics are usually cut to some configuration (Fig. 10.9) and then oriented to the position for sewing or joining.

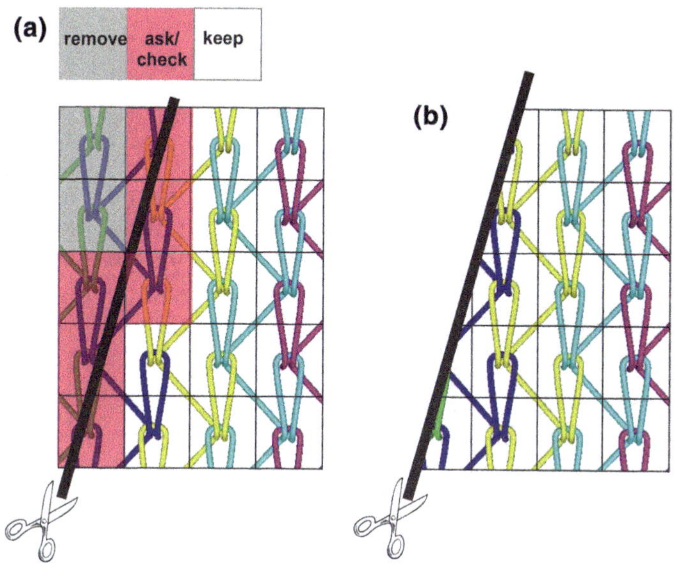

Fig. 10.10 Cutting methods **a** complete cells with the information about the yarn coordinates are removed **b** the yarns directly are cut at some place

From a material point of view, the production of parts with the end form is more productive. This is possible in the weft knitting, but such shape production process is connected with a lot of additional steps of transferring stitches. Therefore, the final efficiency and economical benefit depends on the price of the yarn material, machine and the salaries in the production country. Due to this, the production of textile surfaces with given shape is rather an exception than the normal case. In the normal case high productivity machines produce materials with large standard width and then the small parts have to be cut following some pattern. If such parts have to be modelled at yarn and filament (or fiber) level, also the cutting process has to be implemented for such structures. In software like TexGen [10] the cutting of cuboid of one fabrics (named rectangular domain) is possible. The main use of cuboid is the performing of homogenisation procedure and calculation of the properties of textile reinforced composites. The cutting is implemented using geometrical algorithms for solid bodies. At a topological level, the information about the yarns is defined only based on their axes. For this cutting two strategies are possible depending on the required accuracy of the final result:

- cutting trough removing complete topological cells (Fig. 10.10a)
- cutting trough removing yarn parts (Fig. 10.10b).

The first approach is simpler and faster but it leads to stepwise boundaries. It also has to be decided what happens with the cells that are crossed from the cutting line or curve, meaning if these have to be removed and when. If the fabric is fine and has

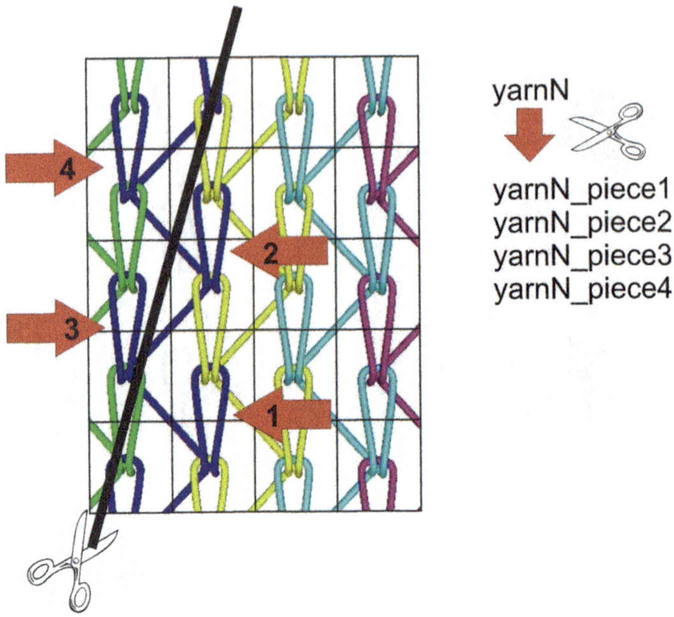

Fig. 10.11 After a fabric is cut into two pieces, the real yarn in the remaining piece is also split into several pieces. The same has to be done with the virtual yarn as well

large numbers of cells, this will not be a problem for the investigation of the behavior at macro level.

The second approach (Fig. 10.10b) of cutting does not use the cells with the topological information. It requires identification of the crossing point of the cutting plane with each yarn axis, creating a new boundary point and removing the remaining parts. This approach leads to clear boundaries, but is complicated regarding the programming, if the cell based data for the structural elements has to be kept. It make sense, if the yarn coordinates of one yarn are kept in a list and not in the cells. This is more efficient and simple for the visualization, but shows disadvantages if the orientation at local places has to be changed after some forming process.

In both situations the cutting process creates several yarn pieces from one yarn, (Fig. 10.11) and this process has to be implemented in the software accordingly.

10.6 Stitching

After the single pieces of fabrics are cut to the required pattern, rotated and placed in the 3D spaces at the proper position, they can be stitched together. The process of stitching can be modelled in different ways, but in this book the complete process

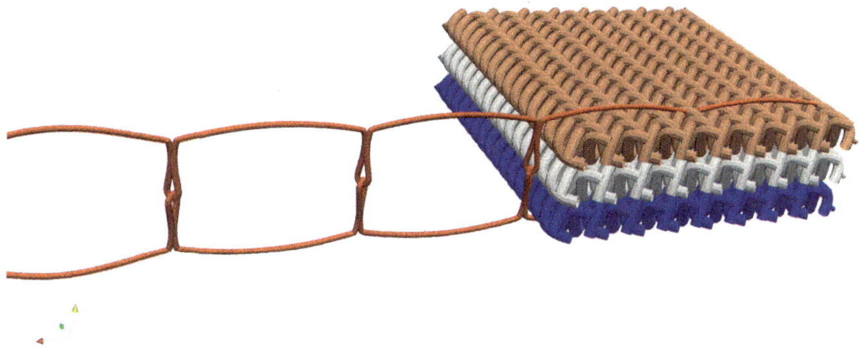

Fig. 10.12 Three fabrics and sewing stitch placed together in the space

simulation (for instance again using explicit FEM) will not be discussed, as it is very time consuming yet. From a topological point of view the first step is based on the marking of the places in the fabrics, where the needle will produce opening [5] and used as basis for the cells of the stitches. The space for the sewing yarn can be opened performing an additional simulation step in which cylinder bodies (the needles) can move the yarns around, but this step can be also be omitted. At the later step, when the sewing yarns are inserted into *yarn* based models of the fabrics and then the contact detection and refinement of the structure is performed, the colliding yarns will rearrange themselves. This step should not be applied at filament level, because in this case there will be additional artificial interlacements between the filaments of the sewing yarns and of the yarns of the textile structure, which do not exists in the reality. Figure 10.12 demonstrates jointed three flat braided fabrics with a topologically sewing stitch without contact correction.

10.7 Conclusion

This chapter summarizes the steps and methods, which are required for the representation of the fabrics pieces in the 3D space for the preparation of their connection with sewing stitches. Different software tools for performing all single steps are available, but the integration of these into one larger tool nowadays is not easy, because they all use different data formats. The completion of such tools, based on the TexMind Suite with exchangeable data, is in progress but not finalized at this time.

References

1. Cebulla, H.: Formgerechte zwei- und dreidimensionale Mehrlagengestricke mit biaxialer Verstärkung: Entwicklung von Maschine, Technologie und Produkten, Dresdner Forschungen Maschinenwesen, vol. 20. TUDpress, Verl. der Wiss, Dresden (2005)
2. Erleben, K., Sporring, J., Henriksen, K., Dohlmann, H.: Physics-Based Animation. Charles River Media Graphics Series, 1st edn. Charles River Media, Hingham, MA (2005)
3. Haupt, M., Lin, H., Cherif, C., Krzywinski, S.: Weft-knitted preforms adapted for crash and 3D applications. J. Fash. Technol. Text. Eng. **s2** (2016). https://doi.org/10.4172/2329-9568.S2-005
4. Helbig, F.U.: Druckelastische 3D-Gewirke: Gestaltungsmerkmale und mechanische Eigenschaften druckelastischer Abstandsgewirke. Südwestdeutscher Verlag für Hochschulschriften (2011). http://www.qucosa.de/fileadmin/data/qucosa/documents/5207/data/Diss.pdf
5. Kyosev, Y., Brinkert, N., Zöll, K.: Strategy for simulating dry preforms for composites connected with sewing stitches. In: Lomov, S., Gorbatikh, L., Swolfs, Y. (eds.) COMPTest 2017 Leuven—8th International Conference on Composites Testing and Model Identification—Presentations. KU LEuven, Leuven (2017)
6. Lomov, S., Bernal, E., Koissin, V., Peeters, T.: Integrated Textile Preprocessor Wisetex, Version 2.5: Computational Models, Methods and Algorithms
7. Lomov, S.V., et.al.: Wisetex (2012)
8. Renkens, W., Kyosev, Y.: Geometry modelling of warp knitted fabrics with 3D form. Text. Res. J. **81** (4), 437–443 (2011). https://doi.org/10.1177/0040517510385171
9. Roye, A., Gries, T.: 3-D textiles for advanced cement based matrix reinforcement. J. Ind. Text. **37**(2), 163–173 (2007/10/01)
10. Sherburn, M.: Geometric and mechanical modelling of textiles. Ph.D. University of Nottingham, Notthingham (July 2007). http://etheses.nottingham.ac.uk/303/1/thesis-final.pdf
11. Trümper, W., Lin, H., Callin, T., Bollengier, Q., Cherif, C., Krzywinski, S.: Recent developments in multi-layer flat knitting technology for waste free production of complex shaped 3D-reinforcing structures for composites. In: IOP Conference Series: Materials Science and Engineering, vol. 141, pp. 012–015 (2016). https://doi.org/10.1088/1757-899X/141/1/012015

Chapter 11
Data Structures for Multiscale Modelling of Flexible Assemblies and Software

11.1 Software Implementation

The algorithms presented in previous chapters have been developed and tested throughout the years within different programming environments. The most effective tool for testing and teaching modelling is and remains Matlab [1], because it combines the power of the modern programming languages including the object oriented programming with the power of the computational tools providing very large sets of numerical methods and visualization tools. The Matlab is especially efficient for the debugging of models and the algorithms. It allows to stop the simulation process at *any* step and to have *convenient* access to all variables, which means that the yarns and their speeds etc. can be checked and also graphically plotted. Due to the licensing issues, especially for teaching, the favorite programming tool was changed to Python in combination with TexGen [2, 3]. Both are open source, free (contrary to Matlab) and accessible for students and researchers for all platforms. TexGen allows quick visualization of the textile structures, so that the developer can concentrate on the textile and geometrical problems. Both environments—Matlab and the combination of Python & Texgen—are excellent for research and education, but have shown some disadvantages at the stage of integration of the software into industrial CAD systems for commercial applications.

TexGen is convenient, but slow for larger structures and the speed becomes unacceptable for the modelling at fiber level for the modern (2017) computers with 16 GB RAM. With the GPL License it obligates the users to distribute the new implemented code as well as open source under GPL License. Keeping the source open and using the community is excellent initiative, but more efficient for packages with larger use where the critical mass of developers is available. There are only a few real software developers of textile and clothing related software worldwide and the majority of these is working for competitive software companies. Therefore, offering some part of the software freely under GPL or any other open source license means that the competitors will have the chance to receive the "know-how" for free, but the reverse effect of having a community with ideas to improve is, is not there.

Y. Kyosev, *Topology-Based Modeling of Textile Structures and Their Joint Assemblies*, https://doi.org/10.1007/978-3-030-02541-0_11

Matlab is a closed source, commercial package, but software developers can redistribute its core if the commercial license is obtained. It can generate as well executables and integrates well in C++ packages, if the additional compiler license is obtained. The licensing fee increase the costs for the final software, which is an economical disadvantage for small budget projects with few final customers. Matlab has its own tools for graphical user interface, but the compatibility of these tools within the versions was not always given and it provides mainly basic elements (buttons, edit boxes), which are enough for large set of applications, but not convenient enough for more specific cases with intensive interaction between the user and the graphical user interface.

After years of testing both ways, it was clear that the current requirements:

- portability between platforms
- long time availability and portability
- Rich of elements, stable and long time compatible Graphical User Interface
- low investment (low cost)
- licensing allowing proprietary use (closed source, commercial applications)
- speed in the computation and visualisation

can be provided if some wide and long time developed modern programming language is used, of which enough libraries are tested and are available. The language C/C++ was chosen, even though it requires higher developer time compared to the use of Matlab or Python. Once the increased time is invested in the definition of classes and libraries, these are reusable for next projects. This was proven throughout the last 6 years. For graphical interface the library wxWidgets was chosen [4]. The 3D visualization is done using the great "The Visualization toolkit" VTK, which provides very good speed and integrates well into the interface of wxWidgets. The matric computations are done both in VTK or within the Eigen C++ library. The implementation of the algorithms now includes several own tools (Fig. 11.1) starting from interface and including a list of several topology generators, which is convenient for the textile user. These are based on the algorithms, described in the previous chapters. All generators produce identical data sets for the textile structures and use one viewer for 3D visualization. Throughout the time, different import and export functions, which allow the geometry data to be stored in suitable formats for another applications, were implemented.

11.2 Modelling Issue—Levels in Textile Structures

For the multiscale modelling of textile assemblies on the yarn and fiber level, at least four structural levels of data structures are required (Fig. 11.2). The stable fibers or the filaments build the basis Level 0 in the mechanical and topographical modelling. These are normally twisted or just plied together into yarns or multifilaments which are Level 1. Very often several (staple or multifilament) yarns are twisted together to another yarn for improving their evenness, getting the required strength or thickness.

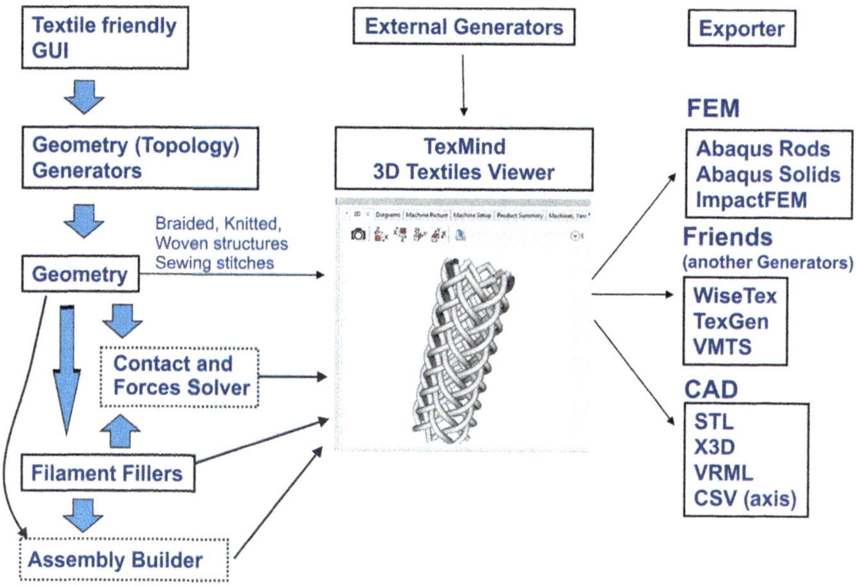

Fig. 11.1 Structure of the implemented software tools for generation, refinement, assembling and export of textile structures and their assemblies for different kind of calculations

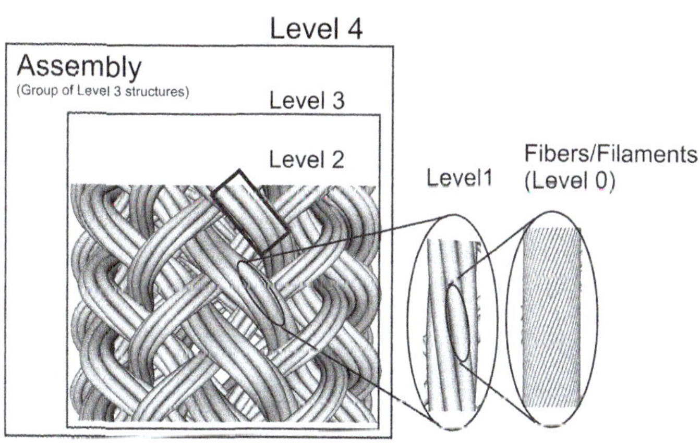

Fig. 11.2 Example of the basic element (Filament) and levels of grouping of the fibrous assemblies in this case for a braided structure [5]

This is represented by Level 2 structure. The yarns build ropes, knitted structures, woven structures or others. These are the structural elements of level 3. In most of the cases, several fabrics pieces are mounted, sewed or connected in another way and made into a final product, which is the assembly. This is presenting level 4 of the hierarchy.

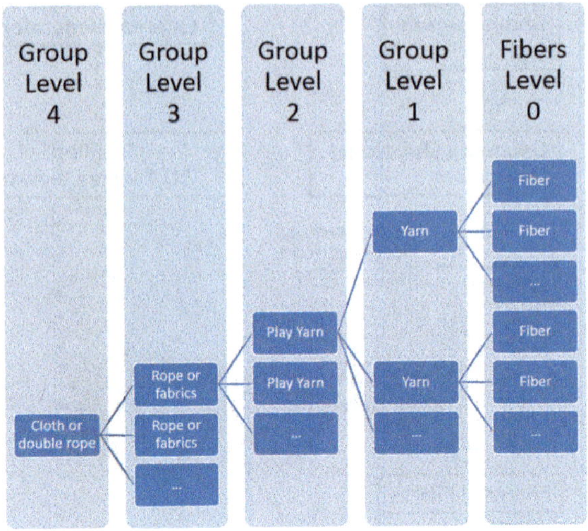

Fig. 11.3 Fibers (or filaments) are basic object of fibrous structures. All higher levels consist of groups of another groups [5]

All these levels needs similar data structures- they all consist of groups of elements from the previous level (Fig. 11.3). It is well visible, that only two main types of virtual objects are required for storage of the data for simulation of textile structures—fibers (fialments) and "groups of fibers" (Fig. 11.4). Each "group of fibers" can be yarn, ply yarn, twisted yarn or braided structure. It is a linear product and it can be characterized from the programming and modelling point of view as a larger and thicker fiber or "micro-fiber". In order for the properties of this macro-fiber to be visualized at the current level, all its properties—diameter, fineness, color and mechanical properties should be defined or calculated. In such recursive definition, in which the main object is "group of fibers", consisting of another "groups of fibers" there is no limitation of the number of the levels of grouping and all kinds of fabrics can be modelled and represented.

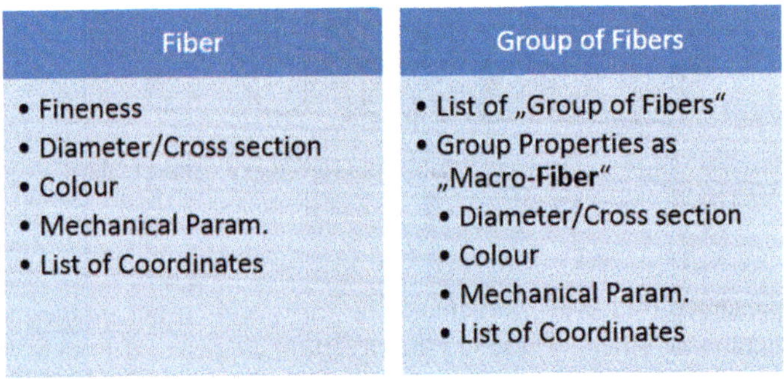

Fig. 11.4 Two required types of objects (classes)—fibers and "group of fibers" and their principle properties. Each group of fibers can also be represented as a macro-fiber (yarn, ply yarn) and because of this the definition of fiber-similar properties as resulting color, diameter and mechanical properties can be useful [5]

11.3 Software Overview—Comparative Analysis of the Available Packages

There are large numbers of investigations and developments of algorithms for modelling of textile structures, but only a small number of these is available or useful after the defence of the thesis of the developer or after the founded research project is finished. There are also large numbers of commercial available packages for the optical design of structures and control of the machines, but not related to the 3D representation of these structures. A list of such packages can be found in [6]. The following Table 11.1 visualizes some of the software packages known or used by the author for the creation of models of textile structures in 3D. Not listed is the software for warp knitting Warp3D, developed by ALC Computertechnik in Aachen, and later

Table 11.1 Overview of ***some*** software packages for modelling of textiles, used or known by the author [5]

Generators	Wisetex suite	TexGen	TexMind Suite	VTMS Virtual Textile Morphology Suite	Multifil	Multiscale designer
Some sources	[7–10]	[2, 11]	[12, 13]	[14, 15]	[16, 17]	[18]
Developer	KU Leuven	Univ. Nottingham	Y. Kyosev—TexMind UG	The University of Dayton	Laboratoire MSSMat—Centrale Paris	Altair, multiscale design systems
License	Proprietary	GPL	Proprietary	Proprietary	Proprietary	Proprietary
Source	Closed	Open	Closed	Closed	Closed	Closed
Structure type						
Woven	Single- and Multilayer woven each yarn individually defined	Single layer, orthogonal, angle interlock, layer to layer, yarn group properties	Basic structures, Each yarn individually defined	Single layer, orthogonal, angle interlock, layer to layer, yarn group properties	Basic structures, (manual generator)	5 Basic structures
Weft knitted	Loop based fine, Tucks need improvement	Scripting or import from TexMind or Wisetex	Plain Loops in flat and tubular structure	(import possible)	(import possible)	–
Warp knitted	Needle stitched only	Scripting or import	Loop3D—several classes structures on single and double needle bed machines	(import possible)	(import possible)	–

(continued)

Table 11.1 (continued)

Generators	Wisetex suite	TexGen	TexMind Suite	VTMS Virtual Textile Morphology Suite	Multifil	Multiscale designer
Braided	Unit cell of biaxial and triaxial braids	Scripting or import from TexMind or WiseTex	Braider—tubular and flat; configurator—custom (Geometry)	Unit cell of biaxial and triaxial, Tubular briads	Manual generation (import possible)	–
Seams	Some stitches	Scripting or import from TexMind or WiseTex	Yes	–	–	–
Yarn level						
Yarn level—Mechanics	Beem—Energy minimization	(scripting possible)	–	Digital chain	Beam	–
Yarn contact	Algorithmically included	–	(Local sections)	Mesh	Beam-beam	Surface-surface
Scripting	–	Python	–	–		–
Imports		WiseTex weave pattern	WiseTex,			Abaqus inp File
Exports	Abaqus, Ansys, WiseTex XML (TexGen)	Step, STL, Abaqus, Voxel Grid, (WiseTex)	Abaqus, Ansys, VTMS, WiseTex, TexGen, ImpactFEM, X3D, STL			
Yarn definition	Initially constant cross section	Variable cross section	Constant Cross section	Constant Cross section		Constant Cross section
Filament level						
Filament visualization	–	–	Yes	yes	Yes	–
Filament data stored	Yes	–	Yes	Yes	Yes	
Filament data used for computations	Yes	–	Yes	Yes	Yes	–
Initial filament distribution in the yarn cross section	–	–	Circular layers, flat layers and arbitrary distribution possible	Number of filaments	Program dependent	–
Structural levels	2(3): Fabrics-Yarn- (Fiber data); In LamTex: 3	3: Layered textile—Textiles—Yarn	4: Textile-4: Yarn Groups-Yarns- Fibers	3: Fabrics-Yarn- Fibers	3 Fabrics-Yarns-Fibers	1: Yarn volume mesh

maintained and developed by its successor Texion; the software for 3D simulation of weft knitting process of the company Shima Seiki, presented during the ITMA 2015 in Milano, and many other commercial weaving design programs, which provide 3D visualiation. All these packages cover only one type of structure (warp or weft knitted or woven) and as pure commercial packages do not publish any information about their model building methods, possibilities and export options.

The table gives a *current* overview about the capabilities of the available programs. All these are being developed and extended continuously. The Multiscale Designer of Altair, acquired of Multiscale Design Systems, LLC, proves that the commercial FEM Software producers already try to integrate textile generators for the calculation of composites in their software. The comparison shows, that

- most of the textile structures are already covered by different packages;
- each package has its own advantages—some are more easy to use and flexible, but not accurate; others cover mechanical calculations but require more data;
- the implemented algorithms in the Texmind software covers areas, which are not covered by the others—braiding, warp knitting and partially weft knitting and stitching.

Most of the packages cover only 3–4 levels of hierarchy, which makes the integration in assemblies difficult.

11.4 Conclusions

This chapter presents an analysis of the required data structures for representing textile products at four scales—fiber, yarn, fabrics, assembly—in a computer software. The analysis results in the conclusion that actually five hierarchical levels have to be included. The basic level is always the fiber or the filament. All other levels can have identical structures, defined as "group of virtual objects". This is a recursive definition, but it is the most simple approach for the programming implementation, because only two basic classes have to be prepared. At the end of the chapter a comparison of the implemented algorithms with similar packages is performed. This comparison demonstrates that the developed and implemented models fill several gaps in the modelling of textiles.

References

1. Mathworks: Matlab, www.mathworks.com (2014)
2. Sherburn, M.: Geometric and mechanical modelling of textiles. Ph.D., University of Nottingham, Notthingham (July 2007). http://etheses.nottingham.ac.uk/303/1/thesis-final.pdf
3. TexGen: http://texgen.sourceforge.net/index.php/Main_Page (2007)
4. Smart, J., Roebling R., Zeitlin, V., et al.: wxwidgets cross platform gui library (2015). http://www.wxwidgets.org/

5. Kyosev, Y.: Geometrical and mechanical modelling of textile structures at fiber and yarn level—software and data structures. Fibres Text. **1**(2017), 3–9 (2017)
6. Kyosev, Y.K.: Simulation of wound packages, woven, braided and knitted structures. In: Veit, D. (ed.) Simulation in Textile Technology, pp. 266–309. Woodhead Publishing Limited (2012). https://doi.org/10.1533/9780857097088.266
7. Lomov, S., et al.: Wisetex (2011). URL http://www.mtm.kuleuven.be/Onderzoek/Composites/Research/meso-macro/textile_composites_map/textile_modelling/textile_modelling_fe
8. Moesen, M., Lomov, S., Verpoest, I.: Modelling of the geometry of weft-knit fabrics. In: Techtextil (Hg.) 2003—TechTextil Symposium 7–10 April 2003 (2003)
9. Moesen, M., Lomov, S.V.: Weftknit (2011). http://liegebeest.studentenweb.org/weftknitEN.html
10. Verpoest, I., Lomov, S.V.: Virtual textile composites software wisetex: integration with micro-mechanical, permeability and structural analysis. Compos. Sci. Technol. **65**(15–16), 2563–2574 (2005)
11. University of Nottingham Textile Composites Research: Texgen (2013). http://texgen.sourceforge.net/index.php/Main_Page
12. Kyosev, Y.K.: Texmind braider, www.texmind.com (2012)
13. Kyosev, Y.K.: Generalized geometric modelling of tubular and flat braided structures with arbitrary floating length and multiple filaments. Text. Res. J. **86**(12), 1270–1279 (2015)
14. Zhou, E., Mollenhauer, D., Larve, E.: A realistic 3-d textile geometric modeling. In: ICCM-17: 17th International Conference on Composite Materials. IOM Communications, London (2009)
15. Zhou, E., Sun, X., Wang, Y.: Multi-chain digital element analysis in textile mechanics. Compos. Sci. Technol. **64**(2), 239–244 (Feb. 2004)
16. Durville, D.: Finite element simulation of textile materials at mesoscopic scale. In: Finite element modelling of textiles and textile composites—St-Petersburg—26–28 September (2007)
17. Durville, D.: Simulation of the mechanical behaviour of woven fabrics at the scale of fibers. Int. J. Mater. Form. **3**(S2), 1241–1251 (2010). https://doi.org/10.1007/s12289-009-0674-7
18. Altair: multiscale design. http://www.altair.com/ (2016)

Part VI
Mechanics After Topology—Application of Topological Based Models for Mechanical Simulations

Chapter 12
Computational Mechanics of the One Dimensional Continuum as Refinement of the Topology Based Models

12.1 Introduction

Textile yarns are one of the most significant representatives of the one dimensional continuum. Their motion is investigated at different levels for almost all textile production processes, but the models from one process are rarely applicable for the other processes. This is due to certain limitations and assumptions in each model of the models. To start the investigation a simple model like the motion of a yarn during unwinding can be used. It presents a unconstrained motion of one dimensional continuum without contacts at the beginning. In the textile structures, the yarns are moving with a lot of constraints, based on contacts to other yarns which cause lateral forces, friction, bending etc. These contacts determine the behaviour of the structure. They can be included in the model at later step, when it is clear, that the yarn motion can be simulated properly.

There are two approaches for creating a mechanical model—force and energy—approaches. These two approaches are equivalent from the point of view of the mechanics, but they differ in the way they form the equilibrium equations of the structures. This equation influences the selection of the numerical method and the programming complexity of the problem. If the energy approach is used—a minimization problem has to be solved. The energy of the system is minimized with several constraints and (a large number of) variables. If the force approach is used, the differential equation of motion has to be integrated. In the following, both of the methods and their advantages and limitations are discussed briefly.

12.2 Energy Minimisation

The mechanical models for textile structures, introduced in [11, 12, 15, 27] etc., differ in the yarn properties used and the definition of the contact zone (one point, two point, contact line etc.). Hart et. al [11, 12] applies mechanical analysis for warp

Y. Kyosev, *Topology-Based Modeling of Textile Structures and Their Joint Assemblies*, https://doi.org/10.1007/978-3-030-02541-0_12

knitted loops too, and demonstrates the influence of the bending rigidity of the yarns over the loop form. The available models differ in the assumptions made about the yarn properties—such as bending, tension, torsion, compression effects and yarn stiffness. Another point in which the models differ is in the treatment of contact between the yarns—in one, two or in more points. An extensive overview of the models can be found in several papers e.g., in [31]. One of the main approaches used to calculate of the loop form—taking into account the mechanical properties—is the minimization of the internal energy of the loops [13]:

$$\min E = \min\left(\int_{o}^{L} E_b + E_r + E_c + E_t ds\right) \tag{12.1}$$

$E_b,\ E_r,\ E_c,\ E_t$ are the bending, torsion, lateral compression and longitudinal tension energy of the single loop. L is the loop (or one half of the loop) length and s is the curvilinear coordinate. All the energy terms can be explained as functions of the local coordinates of the parametric defined yarn axis $z = z(s)$. The contact between the yarns is also included through the lateral compression term

$$E_c = C \cdot g(r) \tag{12.2}$$

where C is the yarn compression rigidity, r distance between the reference yarn to another in the contact area:

$$r = z(s) - \tilde{z}(s) \tag{12.3}$$

Here $g(r)$ is defined semi empirically and has to be able to represent the differences in the behaviour of the different yarns from (near) non-compressible yarns like monofilaments to high compressible yarns like PAN bulk yarns. The continuity conditions apply as an additional constraint:

$$\dot{z}_1^2 + \dot{z}_2^2 + \dot{z}_3^2 = 1 \tag{12.4}$$

In order to find three independent variables (since z_i is connected through 12.4), new variables are introduced and often defined as two rotations (one in plane, one perpendicular to the plane) and the twist rate. More detailed descriptions about the numerical approaches implementing the Lagrange multipliers and the modified Hamiltonian can be found in [13, 31] and several other works. This method demonstrates very good results if applied for single bar structures [11, 12]. This approach is implemented as well in the woven structures modeller Wisetex [29], where the stable state of the yarns can be reached relatively easy. Here, the open problem lies in the rapidly increasing computational complexity if large and more complex structures have to be modelled. In this case, the discretisation and the solution of the equations for small pieces—finite elements, can be the solution. Successful implementation of the energy approach in its implementation as implicit finite element method is reported by Durville [5, 6] etc. His works have proven that this approach also works for general textile structures, but currently is connected with very large computational time. Due to license issues the implementation "Multifil" has restricted access.

12.3 Force Equilibrium Approach

In earlier years the force equilibrium approach was used mainly in the yarn dynamics applications such as the yarn balloon. The modeling of the yarn balloon is the object of a large number of mathematical and mechanical investigations. The different equations of balloon's stable motion are investigated in [1, 8, 9]. The differences in the stationary and transient equations are well described in [28] and in [35]. They also present a transient solution of the balloon equation. The detailed study for unwinding machines and influences of some factors like form of the boundary conditions are also presented in [19] and the following series of works [17, 18] etc. All of these works are based on the solution of the system of partial differential equations with use of finite differences or relaxation technique and shooting method for boundary value problems. They use separate systems of equations for each zone of the yarn balloon and the solutions are synchronized through the common conditions of the boundaries between the zones. Because of this separation, the unique phenomenons in the areas can be investigated in depth, but the model can not be applied for wider applications with more or different zones. With the increasing computational power more applications use methods, in which the discretization along the curve coordinate s is done as a physical model—presenting the yarn as a mass-spring system. These methods provide relatively simple systems of equations, but need more computational steps. In the last years [3, 33, 34] presented several works using mass-spring systems as representations of the yarn and also presented several practical investigations, but the exact mathematical formulation and the numerical method is not well described. For knitted structure mass-spring model is applied in [23]. There are also several works about marine cables based on rigid chains models using the commercial software ADAMS. Basically, the idea is the same as this in the molecular dynamic [20]. These models allow implementations for more general geometrical conditions which is important for industrial applications. In the current work a model of the unwinding process is presented in which the yarn is modeled as a mass-spring system. The equations of motion of the complete yarn length are integrated during the simulation, including the yarn part on the bobbin, the free yarn length in the balloon and the yarn part after the guide. Two integration methods are tested and the maximal stable step size is given.

12.3.1 Continuous Model

The textile yarns are presented as one dimensional continuum. The common used in this case is the Lagrange description of the equation of motion. The independent curve coordinate s is among the yarn axis (12.1) and is moving together with the yarn. For general cases of yarns with linear mass density ρ, yarn cross section A and bending rigidity B, the equation of motion based on the Newton's law is:

$$\rho A\frac{\partial^2 \mathbf{r}}{\partial t^2} = \frac{\partial}{\partial s}(T\frac{\partial \mathbf{r}}{\partial s}) - B\frac{\partial^4 \mathbf{r}}{\partial s^4} + \rho A\mathbf{g} + \mathbf{q} \tag{12.5}$$

where t is the time, $\mathbf{g}$ the earth acceleration, T is the yarn tension and $\mathbf{q}$ is the resulting vector of the external (except the earth acceleration) forces. The extensibility condition is given by

$$\frac{\partial \mathbf{r}}{\partial s} \bullet \frac{\partial \mathbf{r}}{\partial s} = (1+\varepsilon)^2 \tag{12.6}$$

where ε is the elongation of the yarn. The material properties under tension can be presented as sum of one linear elastic part and one more general nonlinear part:

$$T = AE\varepsilon + f(\varepsilon, \frac{\partial \varepsilon}{\partial t}) \tag{12.7}$$

By given boundary and initial conditions the yarn coordinates r and and yarn tension T can be calculated by solving the system 12.5, 12.6 and 12.7, for the each time step t. By assuming $\varepsilon = 0$ the force and elongation can be eliminated and the equations can be solved faster, as reported in several computer graphics papers for inextensible hair simulation for instance. The system presented above has one disadvantage in this form—if modelling of local effects such as yarn guides, contact with machine parts etc, is required, the equation motion for each part and additionally equations for the relations in the contacting areas, which are boundaries between the parts have to be developed and solved. Then, the system 12.5, 12.6 and 12.7 and suitable boundary conditions have to be built separately for each part, but solved together for all yarn parts. This is not suitable for building a more general library for yarn motion calculations. Because of this, a discretized presentation of the yarns makes more sense for the use in textile structures.

12.3.2 Discretised Model

The discrete model can be obtained by presenting the derivatives to s as a discrete function. It also has an own mechanical representation Fig. 12.1b. Let the yarn be divided into yarn segments with initial equal length L. All forces over each piece

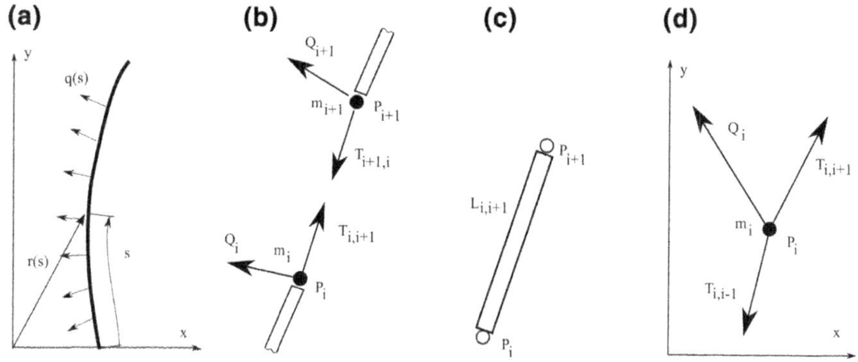

Fig. 12.1 One dimensional continuum (textile yarn) (**a**), discretised on nodes and elements [22]

are concentrated over the nodes P_i between segments. Each node will have mass $m_i = L \cdot \rho$ and the resulting nodal force $\mathbf{F}_i$ is applied.

Equation of motion of single mass point can be given from the second Newton law as:

$$m_i \frac{d^2\mathbf{r}_i}{dt^2} = \mathbf{F}_i \tag{12.8}$$

where the resulting force $\mathbf{F}_i$ at node i is calculated as vector sum of the forces from both neighbour segments $\mathbf{F}_{ei}^{i,i-1}$ and $\mathbf{F}_{ei}^{i,i+1}$:

$$\mathbf{F}_i = \mathbf{T}_{i,i-1} + \mathbf{T}_{i,i+1} + \mathbf{Q}_i \tag{12.9}$$

where $i = 1 - N$ is the node number, $\mathbf{T_{i,j}}$ is the the internal forces in segments $i - j$, and $\mathbf{Q_i}$ are the nodal external forces. Here $T > 0$, as the yarns can transmit only tension, in this case the initial partial differential equation 12.5 is hyperbolic. If becomes $T <= 0$, the PDE type changes and the mechanical instability causes singularity of the equation of the motion. In case of extensible yarns, the extensibility equation in this case is defined at the segment level from the distance between two nodes

$$\varepsilon_{i,i+1} = \frac{|P_i P_{i+1}| - L_0}{L_0} \tag{12.10}$$

and the material properties are defined as discretized form of the Eq. (12.7):

$$T_{i,i+1} = AE\varepsilon_{i,i+1} + f(\varepsilon_{i,i+1}, \frac{\partial \varepsilon_{i,i+1}}{\partial t}) \tag{12.11}$$

Equation (12.9), together with (12.10) and (12.11) build a system from $3 \times N$ ordinary differential equations for the all N nodes in $3D$ space. This can be solved using various numerical methods for ODE, but only few of them are suitable for the practical use.

12.3.3 Dimensionless Discretized Model

Applying the Buckingams Theorem [4], the variables can be reduced and the system can be written in dimensionless form. Let ρ be the linear density of the yarn, r unwinding radius (see Fig. 12.2), V—yarn velocity, T—the time for unwinding of the yarn in one full revolution, L—element length, ω the angular velocity of the unwinding point and the dimensionless time $\bar{t} = \frac{t}{T}$, where $T = \frac{2\pi r}{V} = \frac{2\pi}{\omega}$. Dimensionless coordinates will be $\bar{x} = \frac{x}{r}$, and the dimensionless force will be according to [35]: $\bar{F} = \frac{F}{\rho V^2} = \frac{F}{\rho \omega^2 R^2}$. By putting these variables in Eq. (12.8) and some rearrangements it can be obtained

$$\frac{d^2\bar{x}}{d\bar{t}^2} = \frac{V^2\, T^2}{L\, r}\, \bar{F} \tag{12.12}$$

Usually in ring-spinning here the variable $\xi = \frac{V}{r \cdot \omega}$ is introduced, where for ring, rotor and centrifugal spinning spinning $\xi \neq 1$. For the case of axial unwinding or all other cases, where the bobbin with the yarn is fixed, not rotating, $\xi = 1$is valid. Additionally the relation between element length L and the unwinding radius r can be noted as $\lambda = \frac{r}{L}$. Applying these relations the final dimensionless equation of motion can be obtained:

$$\frac{d^2\bar{x}}{d\bar{t}^2} = 4\,\pi^2\,\xi\,\lambda\,\bar{F} \tag{12.13}$$

For the boundary condition the velocity is needed in dimensionless form, too. Using the introduced variables it can be obtained that $\bar{v} = v\,\frac{T}{r} = v\,\frac{2\pi}{V}$. The elongation ε is dimensionless variable, but during the force calculation an appropriate elasticity module has to be chosen. If both sides of (12.11) are divided by $\rho\, V^2$, for the linear material it can be obtained:

$$\bar{T}_{i,i+1} = \frac{AE}{\rho V^2}\varepsilon_{i,i+1} \tag{12.14}$$

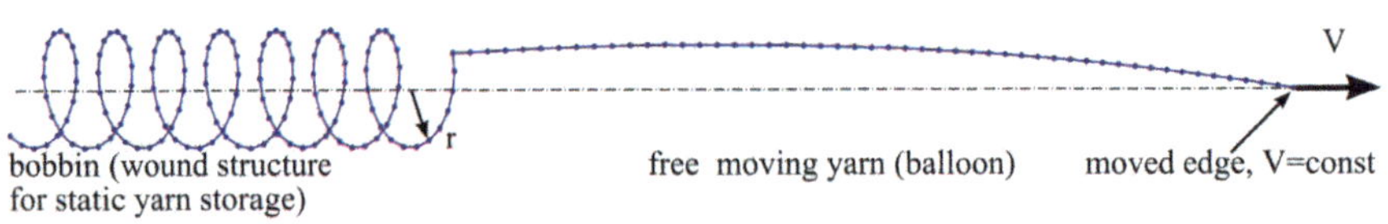

Fig. 12.2 Geometry of the problem [22]

12.3.4 Boundary Conditions

First point of the chain—node $i = 1$ has a fixed velocity:

$$v_{1,x} = 0 \qquad v_{1,y} = 0 \qquad v_{1,z} = V = const \tag{12.15}$$

All other points have initial state, as presented on Fig. 12.2, but are not fixed.

12.3.5 Numerical Integration

The accelerations in time step j are calculated according to Eq. (12.8) as

$$\mathbf{a}^{(j)} = \frac{d^2\mathbf{r}^{(j)}}{dt^2} = \frac{\mathbf{F}^{(j)}}{m_i} \tag{12.16}$$

12.3.5.1 Critical Step Size

For the explicit integration methods the stability limit is defined in terms of maximal time-step Δt depending on the highest frequency in the system ω_{max} [10, 30]:

$$\Delta t_{stable} = \frac{L_0}{c} \qquad c = \sqrt{\frac{E}{\rho_m}} \qquad \rho_m = \frac{\rho}{A} \quad \Rightarrow \quad \Delta t_{stable} = L_0\sqrt{\frac{\rho}{E\,A}}$$

For the mass-spring system the critical step-size will be the same since $\Delta t_{stable} = \frac{2}{\omega_{max}}$, $\omega_{max} = \sqrt{\frac{k}{m}}$, $k = \frac{AE}{L}$ and $m = \rho L$. A more accurate estimation of the maximal time step, taking into account that the one dimensional continuum is moving among three dimensional space, is described in [32] for propagation problems the famous Courant-Friedrichs-Lewy stability criterion for three-dimensional problems is applicable: $\Delta t \leq \frac{\Delta x}{\sqrt{3}|V|}$.

12.3.5.2 Verlet

The Verlet—algorithm uses explicit central difference for discretization of the time derivatives.

$$\mathbf{r}^{(j+1)} = \mathbf{r}^{(j)} + \mathbf{v}^{(j)}\,\Delta t + \frac{1}{2}\,\mathbf{a}^{(j)}\,\Delta t^2 \tag{12.17}$$

$$\mathbf{v}^{(j+1)} = \mathbf{v}^{(j)} + \frac{1}{2}\left(\mathbf{a}^{(j)} + \mathbf{a}^{(j+1)}\right)\Delta t \tag{12.18}$$

12.3.5.3 Leapfrog

The Leap-frog algorithm calculates the velocities for time of a half step.

$$\mathbf{r}^{(j+1)} = \mathbf{r}^{(j)} + \mathbf{v}^{(j+\frac{1}{2}\Delta t)}\,\Delta t \tag{12.19}$$

$$\mathbf{v}^{(j+\frac{1}{2}\Delta t)} = \mathbf{v}^{(j-\frac{1}{2}\Delta t)} + \mathbf{a}^{(j)}\,\Delta t \tag{12.20}$$

12.3.6 Numerical Results

The calculated tension force of the second particle of yarn using time step $\Delta t = 0.004$ and $\Delta t = 0.002$ is presented in Fig. 12.3. The relative error for each of the time steps for the case of $\Delta t = 0.004$ is presented in the same figure and they do not exceed 1%. Actually, the bigger error occurs during the initial time steps, where a transient process is running. This is caused by setting the initial velocity of all points of the yarn except the first to zero. After this transient stage a very smooth stage with almost constant yarn tension follows.

It is well known from the theory that the equation of free motions of the yarn has more than one solutions. On the praxis, they are presented as "single balloon", "double yarn balloon", "tripple" etc. During the time $\bar{t} = 0$–1 there occurs single balloon. The next time step the free yarn length is already greater and more energy is needed to stabilize the single balloon. The system does not have this energy and switches itself into double balloon, which corresponds to the switching of the equation to another solution. This change is well recognizable in Fig. 12.4 at time $\bar{t} = 1$–2. In Fig. 12.3 it is not recognizable, because the selected node is quite far from the region with motion changes. After time $\bar{t} = 2$. After time $\bar{t} = 2$ there is an additional switch

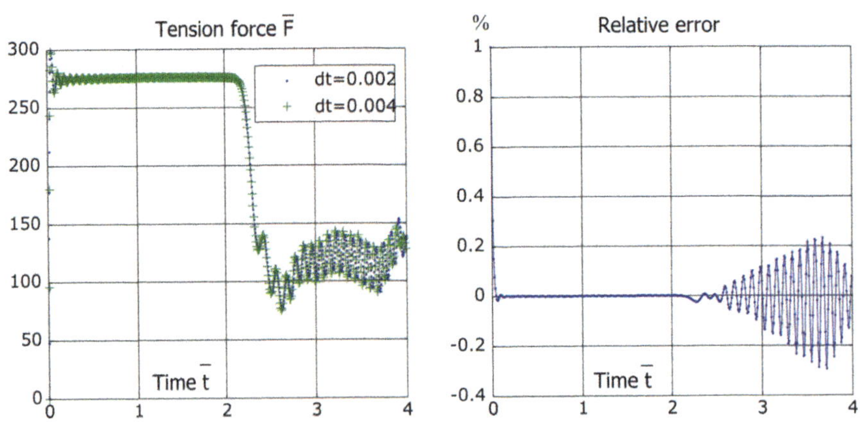

Fig. 12.3 Tension force in the second node of the yarn during the first 4 unwinding cycles, calculated with two time-steps and relative error [%], [22]

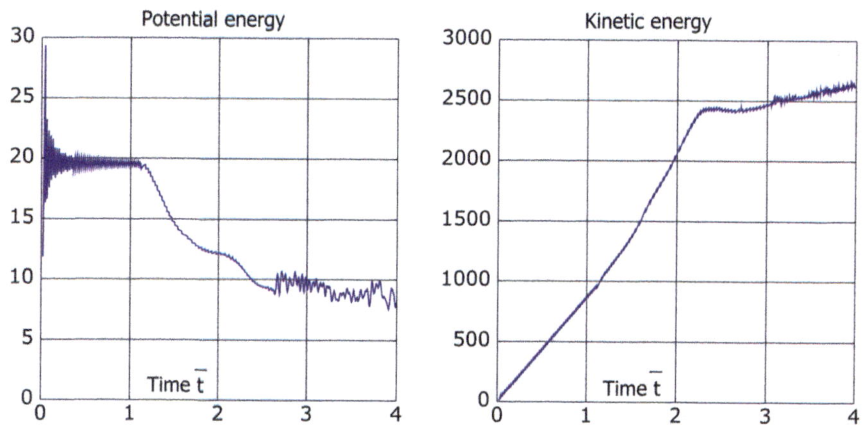

Fig. 12.4 Mean potential and kinetic energy per moving node of the system [22]

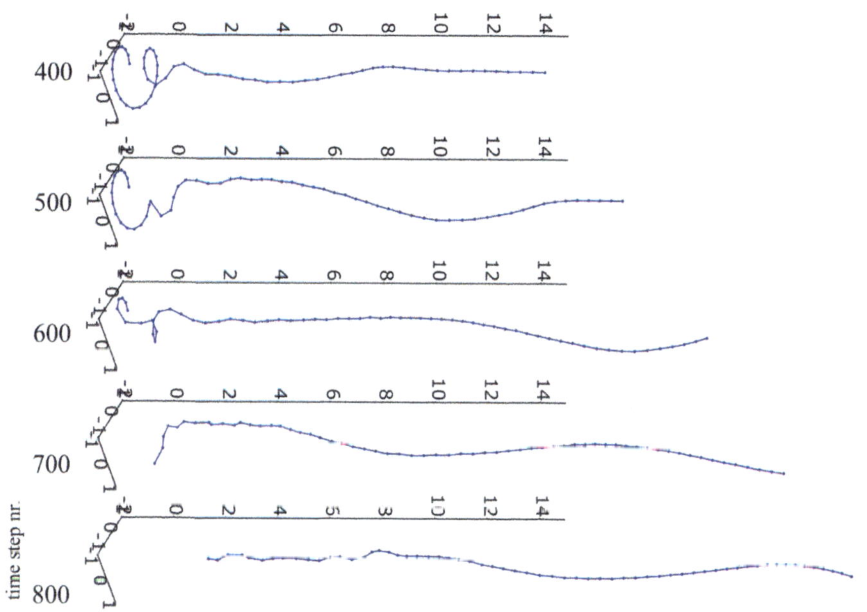

Fig. 12.5 Yarn geometry during unwinding each 100 integration steps, [22]

to triple balloon. After that the system is oscillating with the risk to loose stability. In almost all practical cases, the yarn is going through special guide elements which limit the free length and do not allow going into the unstable state. Therefore, the current algorithm can be applied successfully for all these problems. It presents the yarn motion in a proper way, also in the case of switching between different solutions, which is not the case for all currently published algorithms. Some yarn geometries are presented in Fig. 12.5. Here it has to be noted that the energy is calculated as mean energy of the system per moving node and not as energy for the whole system. This is because the length (and the mass) of the moving yarn increase and thus the the absolute energy of the system increase too.

12.4 Force Equilibrium Approach for Textile Structures

The demonstrated method for the modelling of the free motion of yarns can also be applied for textile structures. For the simulation of the stress relaxation of the structures and some tensile tests the inertia forces are not relevant, as the mass of the yarn is negligible small. As the model actually presents a mass-spring system, it is normal that it starts to swing during the simulation. This can be reduced by including numerical damping.

12.4.1 Damping

In order to avoid oscillations of the mass-spring system, different damping forces can be included. The most simple realization of the damping effect is its implementation into the Verlet equation 12.17 with damping factor f_D, where $0 \leq f_D << 1$ [22]. Equation 12.17 in this case will be:

$$\mathbf{r}^{j+1} = (2 - f_D)\,\mathbf{r}^j - (1 - f_D)\,\mathbf{r}^{j-1} + \mathbf{a}^j \cdot (\Delta t)^2 \tag{12.21}$$

The system integrates until the changes in the coordinates will be smaller then a given error

$$\left|\mathbf{r}^{j+1} - \mathbf{r}^j\right| < \varepsilon \tag{12.22}$$

12.4.2 Bending

The bending is modelled using linear beam theory model:

$$\mathbf{M}_i = K_b \cdot \kappa \cdot \mathbf{n}_{i-1,i,i+1} \tag{12.23}$$

where the curvature could be expressed during simplified linear relation from the bending angle as $\kappa_i = \frac{1}{\rho} = \frac{\theta_i}{L_{i,i-1} + L_{i,i+1}}$ and $\mathbf{n}_{i-1,i,i+1}$ is the normal of the plane, where the links between i − 1, i and i + 1 are located. The bending moment is applied to the nodes through its equivalent forces.

Figure 12.6 demonstrates an example application of the Netwon's law with the damping, bending force and numerical integration using the Verlet algorithm

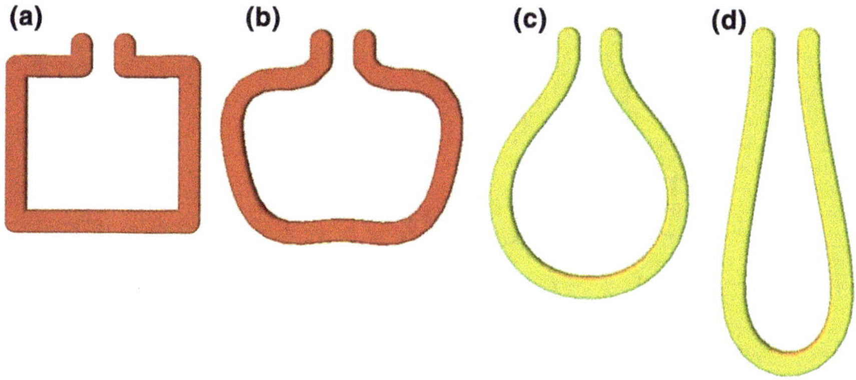

Fig. 12.6 Simulation of a single yarn under impact of a pure bending forces **a** initial configuration; **b** and **c** intermediate states, **d** final state. The damping was set to 0.01. Reprinted from X. Chen (Ed.) "Modelling and predicting textile behaviour", Y. Kyosev, W. Renkens, "Modelling and visualization of knitted fabrics", p. 225–262. Copyright (2010) with permission from Elsevier [24]

starting from simple topological description of a loop based on 8 points. The figures demonstrates the development of the loop form during the relaxation.

12.5 Contact Treatment

The most intensive computational task in the modelling of structures is the contact identification, resolving and treatment. The theory about the Finite Element formulation of the problems is the object of several works. The book [36] covers the most important aspects of its FEM realization, the habilitation work of Konyukhov [16] deals with geometrically exact theory for contact interactions. The implementation in the explicit FEM code is discussed in [2]. Probably less exact, but faster computation proposes are implemented in the software for computer graphics, as explained for instance in [7]. Important from the scientific point of view (not from the efficiency) is the way of how the contact force is computed. For knitted structures Kaldor et al. [14] found out that the definition of the contact force $\mathbf{C}_i$, defined in the following way, leads to better results and stable algorithm:

$$C_i = k_{contact} \cdot f\left(\frac{|\mathbf{x}_i - \mathbf{x}_j|}{2r}\right) \cdot \mathbf{n}_{ij} \tag{12.24}$$

where the function

$$f(d) = \begin{cases} \frac{1}{d^2} + d^2 - 2 & d < 1 \\ 0 & otherweise \end{cases} \tag{12.25}$$

allows better treatment of the contact forces.

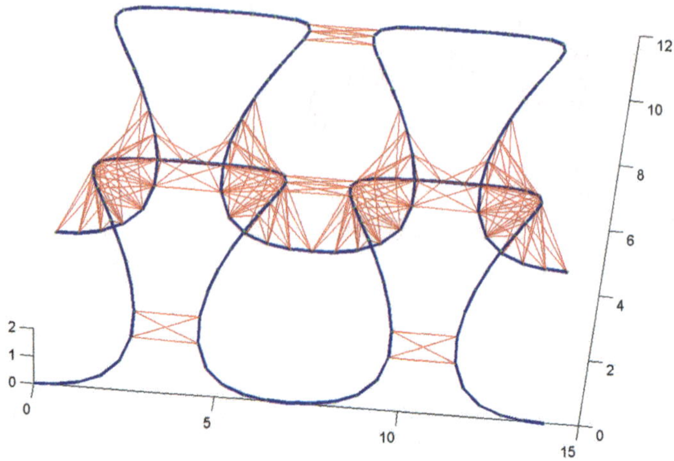

Fig. 12.7 Weft knitted loop geometry with directions of acting contact forces. Reprinted with permission from Yordan Kyosev, Y. Angelova, R. Kovar, (2005) "3D Modeling of Plain Weft Knitted Structures of Compressible Yarn", Research Journal of Textile and Apparel, Vol. 9 Issue: 1, pp.88–97, https://doi.org/10.1108/RJTA-09-01-2005-B009 (C) Emerald Publishing Limited all rights reserved [23]

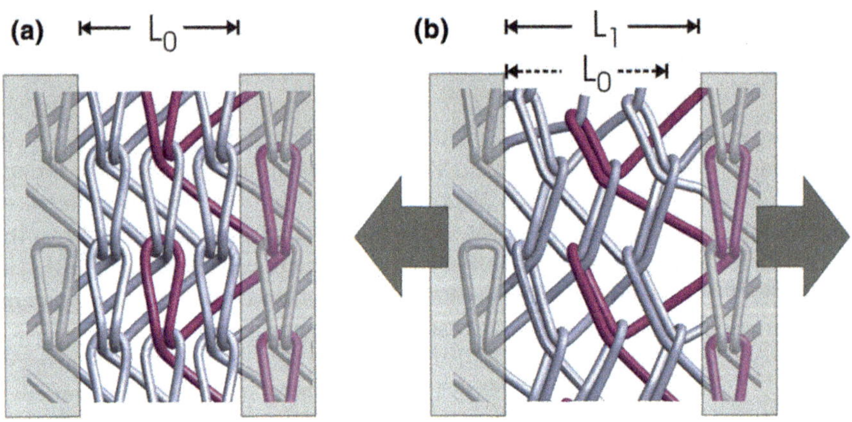

Fig. 12.8 Simulation of a relaxation with deformation in horizontal direction, **a** initial state, **b** final state [25, 26]

The main part of this algorithm (with different contact forces calculation) was implemented in Matlab and tested in the years 2004–2005 as reported in [21]. Figure 12.7 demonstrates the axes of four weft knitted loops (blue) and the contact directions of the identified contact forces (in red) actually, after longer computational time with the available computers in that time the computed displacement were so small, that were optically nor recognizable.

Improved implementation and tests conducted a few years later demonstrate that the method is able to simulated textile structures [26]. Figure 12.8 demonstrates warp

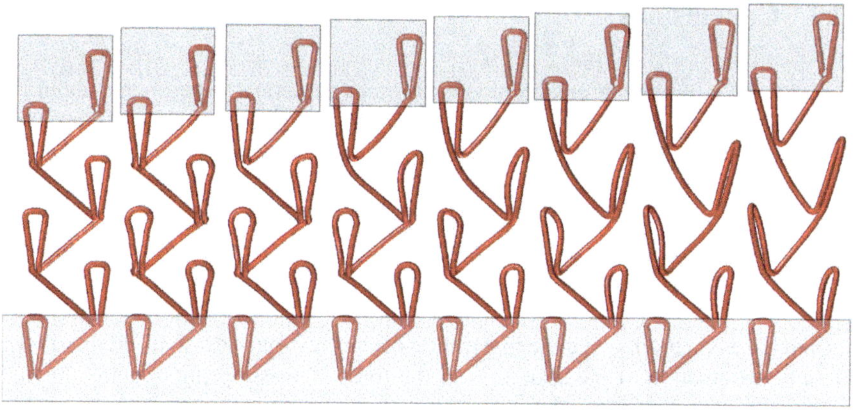

Fig. 12.9 Deformation of the loops in one yarn during tensile test simulation with loading in vertical direction, [26]

knitted structure, generated initially using topologically based method, before and after simulation of horizontal deformation. Figure 12.9 visualizes the changes in one yarn during similar test, but with loading in vertical direction. The calculation of these results has taken less then 10 min per image, but with switched off friction.

12.6 Friction

The friction force has a significant influence on the force-elongation behaviour of knitted structures and causes in large extend the hysteresis in their unloading curves. For woven and braided structures it has secondary influence. The implementation of the friction between the yarns has to be done after the calculation of the contact force. A friction force calculation was included in the implementation [26], including:

- check and calculate the normal contact force
- calculate the resulting friction force
- check if resulting sum of all nodal and external (except the friction) forces is higher than the calculated friction force—if yes, apply the difference; if not—do not apply it.

Because the algorithm has no implemented optimized contact search method (like AABB threes or other) the numerical experiments have taken about 8–10 times longer for the simulation compared to the simulation of the same real time interval, but without contact force. Therefore, this investigation and implementation were kept for the future developments.

12.7 Conclusions

This chapter presents various considerations and numerical experiments about the choose of modelling approach and numerical methods for simulation of the mechanical behaviour of textile structures. These methods can be used for refinement of topologically generated initial configurations with consideration of the mechanical properties of the yarns. The energy based methods are reported to be successfully implemented using Finite Element implementation, but their direct application for structures is not suitable for general algorithms for textile structures, because of the large number of constraints due to interactions between the yarns. The force equilibrium based method, using the integration of the Newton's law in discredited form allow representation of the structures in the form of mass-spring systems. In such a case, the implementation of the contact and other external conditions are applied as external forces and are more elegant and convenient for software implementation. The system is applied for free motions of yarn during unwinding and after that extended for knitted loops but can be applied for any kind of structures. This chapter covers only the author's experiments in this direction. Implemented methods using another (external software) are subject of the next chapter.

References

1. Batra, S.K., Ghosh, T., Zeidman, M.: Integrated approach to dynamic analysis of the ring spinning process, part ii: with air drag. Text. Res. J. **59**, 416 (1989)
2. Belytschko, T., Liu, W., Moran, B., Elkhodary, K.: Nonlinear Finite Elements for Continua and Structures, 2nd edn. Wiley, Chichester (2014)
3. Berger, R.: Instationäre bewegung und stabilitätsverhalten eindimensionaler kontinua. In: Fortschrittberichte VDI, 18. VDI Verlag (1996)
4. Buckingham, E.: On physically similar systems; illustrations of the use of dimensional equations. Phys. Rev. **4**(4), 345–376 (1914). https://doi.org/10.1103/PhysRev.4.345
5. Durville, D.: Finite element simulation of textile materials at mesoscopic scale. In: Finite Element Modelling of Textiles and Textile Composites—St-Petersburg—26–28 September (2007)
6. Durville, D.: Simulation of the mechanical behaviour of woven fabrics at the scale of fibers. Int. J. Mater. Form. **3**(S2), 1241–1251 (2010). https://doi.org/10.1007/s12289-009-0674-7
7. Erleben, K., Sporring, J., Henriksen, K., Dohlmann, H.: Physics-Based Animation, 1st edn. Charles River Media graphics series. Charles River Media, Hingham, MA (2005)
8. Fraser, W.B.: Ring spinning: modelling yarn balloons mathematically. Text. Horiz. **16**(2), 37–39 (1996)
9. Ghosh, T., Batra, S., Murthy, A.: Dynamic analysis of yarn unwinding from cylindrical packages. part i: parametric studies of the two-region problem. Text. Res. J. textbf71(9), 771–778 (2001)
10. Hallquist, J.: Ls-dyna theoretical manual. www.dynamore.de
11. Hart, K., de Jong, S., Postle, R.: Analysis of the single bar warp knitted structure using an energy minimization technique: part i: theoretical development. Text. Res. J. **55**(8), 489–498 (1985). https://doi.org/10.1177/004051758505500807. http://trj.sagepub.com/cgi/content/abstract/55/8/489

12. Hart, K., de Jong, S., Postle, R.: Analysis of the single bar warp knitted structure using an energy minimization technique: part ii: results and comparison with woven and weft knitted analysis. Text. Res. J. **55**(9), 530–539 (1985). https://doi.org/10.1177/004051758505500903. http://trj.sagepub.com/cgi/content/abstract/55/9/530
13. de Jong, S., Postle, R.: A general energy analysis of fabric mechanics using optimal control theory. Text. Res. J. **48**(3), 127–135 (1978). http://dx.doi.org/10.1177/004051757804800302
14. Kaldor, J., James, D.L., Marschner, S.: Simulating knitted cloth at the yarn level. In: Proceedings of SIGGRAPH 2008. Held in Los Angeles, California, August 2008 (2008)
15. Konopasek, M.: Classical elastica theory and its generalizations: the mechanics of flexible fibre assemblies. pp. 255–274. Sijthoff and Noordhoff (1980)
16. Konyukhov, A.: Geometrically Exact Theory for Contact Interactions. KIT Scientific Publishing, Karlsruhe (2011)
17. Kothari, V., Leaf, G.: Unwinding of yarns from packages. part ii. unwinding from cylindrical packages. J. Text. Inst. **70(3)**, 96 (1979)
18. Kothari, V., Leaf, G.: Unwinding of yarns from packages. part iv. two-for-one twisting: the storage disc and balloon. J. Text. Inst. **70(5)**, 184 (1979)
19. Kothari, V., Leaf, G.: Unwinding of yarns from packages. part iii. unwinding from conical packages. J. Text. Inst. **70(5)**, 172 (1979)
20. Krivtzov, A.M., Krivtzova, N.V.: Particle method and its application in the mechanics of elastic solids (in Russian). Dalnevostochnui matematicheskii Journal DVO RAN **3**(2), 254–276 (2002)
21. Kyosev, Y., Angelova, Y., Kovar, R.: 3d modelling of plain weft knitted structures from compressible yarn. Res. J. Text. Appar. **9**, 88–97 (2005). Hong Kong
22. Kyosev, Y., Todorov, M.: Computational model of 1d continuum motion. case of textile yarn unwinding without air resistance. In: Lecture Notes in Computer Science (2007)
23. Kyosev, Y.K., Angelova Y., Kovar, R.: 3d modelling of plain weft knitted structures from compressible yarn. Res. J. Text. Appar. **9**(1), 88–97 (2005). Hong Kong
24. Kyosev, Y.K., Renkens, W.: Modelling and visualization of knitted fabrics. In: Chen, X. (ed.) Modelling and Predicting Textile Behaviour, pp. 225–262. Woodhead Publishing and In association with the Textile Institute and CRC Press, Cambridge and Boca Raton and Fla (2010)
25. Kyosev, Y.K., Renkens, W.: Computational mechanics of knitted structures—critical overview. In: Eberhardsteiner, J., et al. (ed.) European Congress on Computational Methods in Applied Sciences and Engineering (ECCOMAS 2012) (2012)
26. Kyosev, Y.K., Renkens, W.: Simulation of the Mechanical Behaviour of Warp Knitted Fabrics (2013)
27. Leaf, G.G.A.: The geometry of a plain knitted loop. J. Text. Inst. **45**, T587–605 (1955)
28. Lisini, G.E.A.: A comparison of stationary and non-stationary mathematical models for the ring-spinning process. J. Text. Inst. **83(4)**, 550–559 (2001)
29. Lomov, S.V., et al.: Wisetex (2012)
30. NN: Getting started with abaqus
31. Postle, R., Carnaby, G., de Jong, S.: The Mechanics of Wool Structures. Ellis Horwood Limited (1988)
32. Press, W.H., Teukolsky, S.A., Vetterling, W.T., Flannery, B.P.: Numerical Recipes in C: The Art of Scientific Computing. Cambridge University Press (1992)
33. Przybyl, K.: Dynamics of yarn in the process of its formation on the ring-spinning machine. Fibres Text. East. Eur. **3**(3), 36–38 (1995)
34. Przybyl, K.: Modelling yarn tension in the process of manufacturing on the ring-spinning machine. Fibres Text. East. Eur. **6**(3), 30–33 (1998)
35. Stump, D.M., Fraser, W.B.: Transient solutions of the ring-spinnign balloon equations. J. Appl. Mech. **June**, 523–528 (1996)
36. Wriggers, P.: Computational Contact Mechanics. Springer (2006)

Chapter 13
Applications of the Topological Generated Models

13.1 Introduction

The 3D geometry of textile structures generated using topological models do not represent the real geometry with high accuracy, but already contain the important information about the orientation of the yarns. In most cases the accuracy of the models can be increased after careful identifying of parameter values of the yarns and the resulting structure [10]. In other cases the properties of the structures can be evaluated without performing relaxation and contact adjustments. The following examples demonstrate applications in which the geometry is generated using the algorithms from the previous chapters and exported to a specific software to perform the required simulations.

13.2 Deformations of Weft Knitted Structures With LS-Dyna and Hyperworks

One of the first trials to apply explicit FEM software, in this case LS-Dyna, was performed for weft knitted structures [8]. The yarn axis is created based on the geometric model Fig. 13.1. The volume is generated by sweeping a circle along the axis within the Ansys, used as preprocessor, and then meshed with solid elements. The deformation of the loops in the contact places is presented in Fig. 13.2 and the evaluation of the stresses within the elements is presented in Fig. 13.3. The diagrams show peaks in the force development because of the very coarse mesh used.

After the evaluation the explicit simulation of the deformation of the weft knitted structures was used for optimization of bullet proof materials. In the master thesis of K. Freier, structures with different loop sizes and different yarn diameter to loop size ratios were investigated numerically [4]. The configuration which absorbs most of the energy of the ballistic particles was chosen for further optimization and application under industrial conditions. Figure 13.4 shows the velocities of the deformed knitted

Y. Kyosev, *Topology-Based Modeling of Textile Structures and Their Joint Assemblies*, https://doi.org/10.1007/978-3-030-02541-0_13

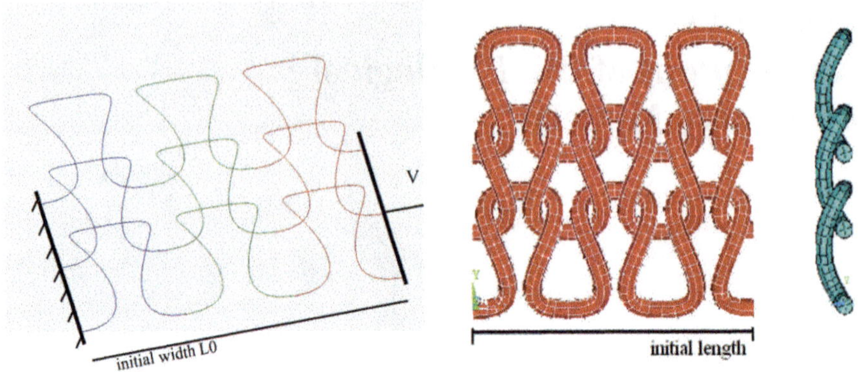

Fig. 13.1 Initial axis of weft knitted structure and generated mesh for 3D solid elements for FEM simulations [8]

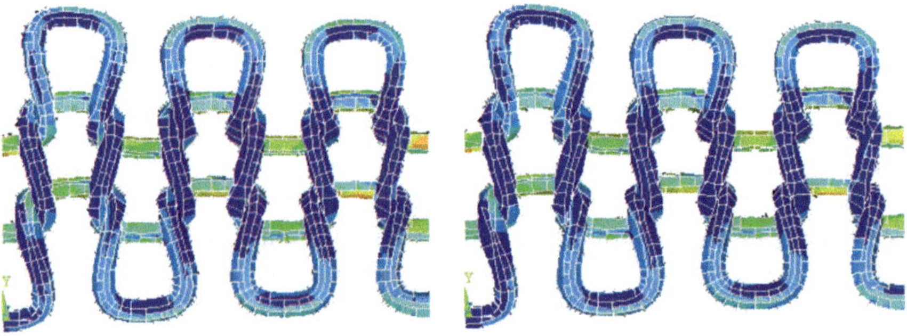

Fig. 13.2 Deformation of the loops considering the friction within Ansys/LS-Dyna [8]

structure and the different energies in the system. The friction between the yarns is considered in the simulation.

13.3 Truss and Beam Finite Elements

For the simulation of textile structures the use of truss or beam elements is effective. These are used efficiently for instance for draping simulations [2]. The truss elements can not transfer bending on the nodes and can be used in tensile loaded structures, in which the bending is not important. An example is the investigation of the tensile behavior of braided ropes. The geometry of the rope is generated using the algorithm of [14] and described in Chap. 2 including its implementation in the TexMind Braider [12]. From there is generated automatically the mesh with trusses and saved in suitable format for the explicit FEM software Impact FEM [3]. Some steps of the tensile deformation and the resulting force until break of the yarns are presented in Fig. 13.5 [9, 13].

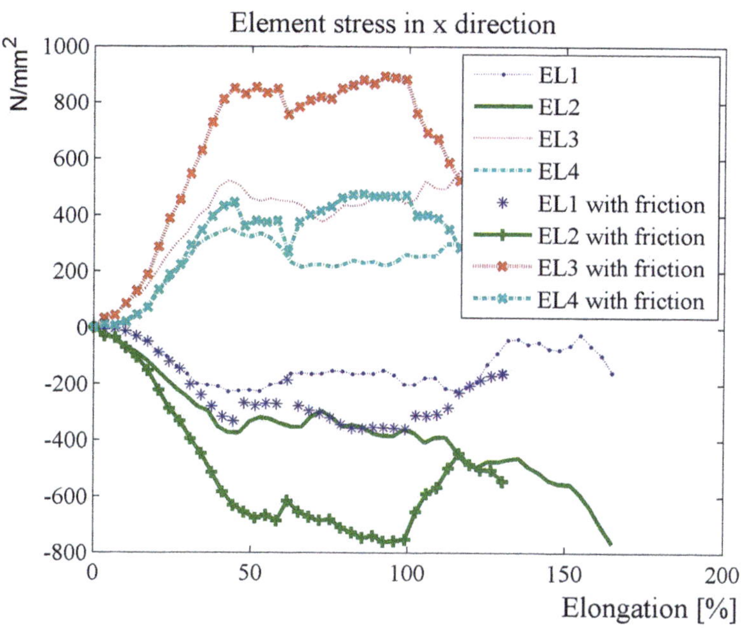

Fig. 13.3 Simulated element stress. Due to the edges and the coarse mesh there are several jumps in the stresses [8]

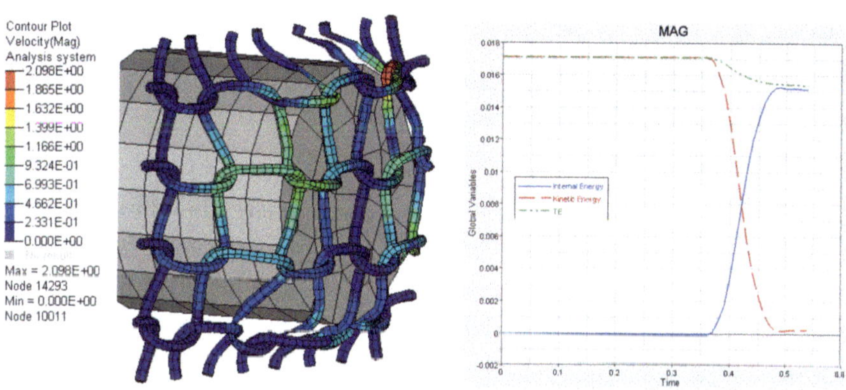

Fig. 13.4 Deformed knitted structure from ballistic particle and plot of the kinetic and potential energy of the system [4]. Knitted structure generated and exported from the TexMind weft knit generator to Hyperworks

The bending of the yarns is important in some application areas, as for instance for the investigation of the compression behavior of spacer warp knitted structures. There are analytical models for the calculation of bending force in pile yarns [5], but they require the parameters and geometry of each pile yarn. Once the warp knitting pattern changes, the pile yarn changes its configuration. With the geometrical model

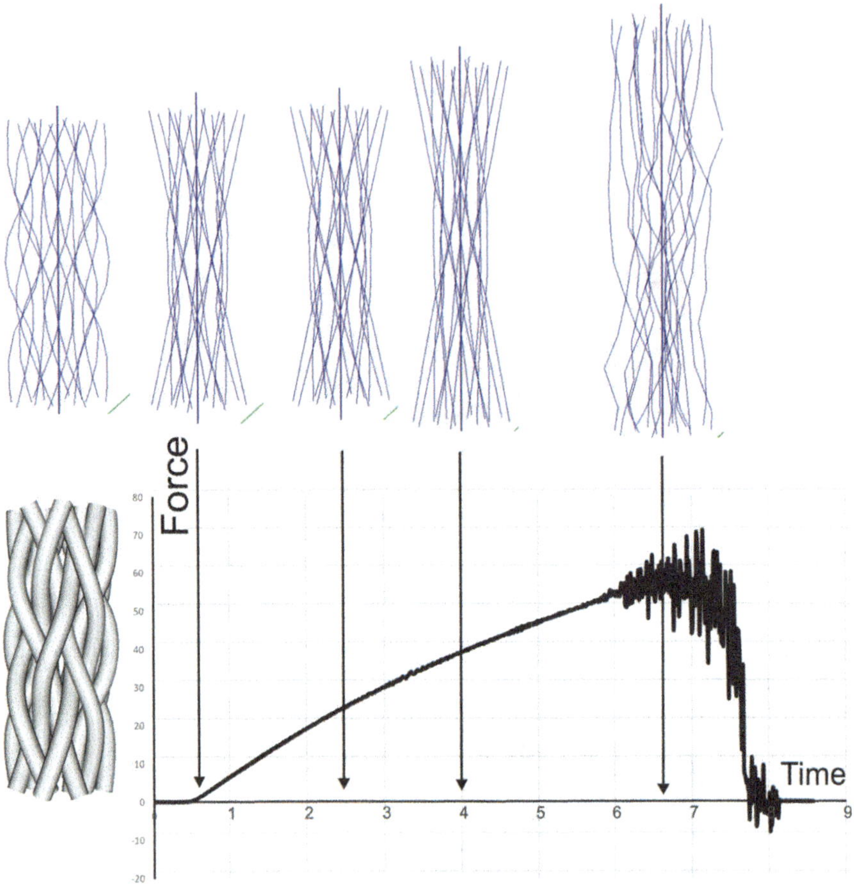

Fig. 13.5 Initial geometry and resulting mesh for explicit calculation. The few geometries show the state of the rope during the loading and at the moment of first yarn breakage [13]

such a geometry can be created parametrically and exported in a suitable format for computations for example as beam based finite element mesh (Fig. 13.6). This is done for one example with the software Ansys and reported in [11].

13.4 Digital Chain Method

The digital chain method [20, 21], is a modified mass-spring model, developed for efficient simulation of fiber based structures. Its implementation in Virtual Textile Morphology Suite (VTMS) uses a data format for the fabrics and yarns, which is covered by the TexMind Viewer. This allows quick exchange of the models. Figure 13.7a demonstrates a braid that is imported in VTMS and which was initially generated

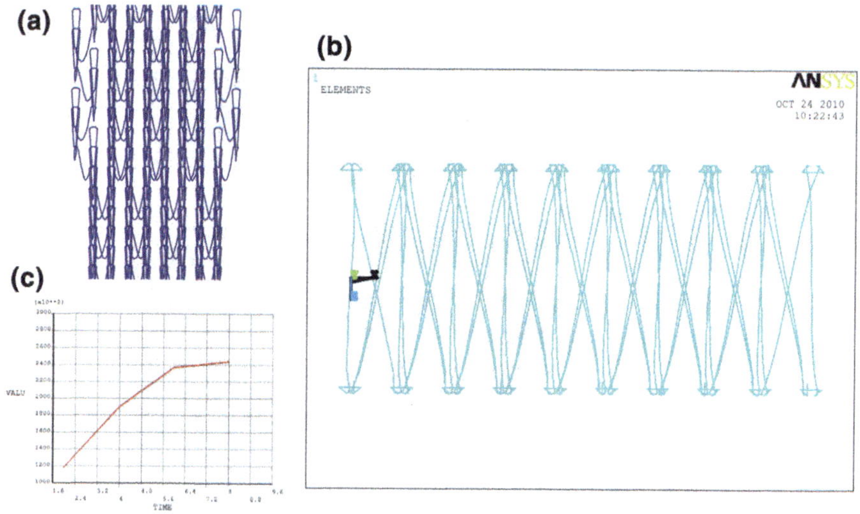

Fig. 13.6 Beam element modelling of spacer warp knitted structure. **a** Geometry, imported in Ansys **b** view through the fabric thickness, **c** simulated displacement—compression curve [11]

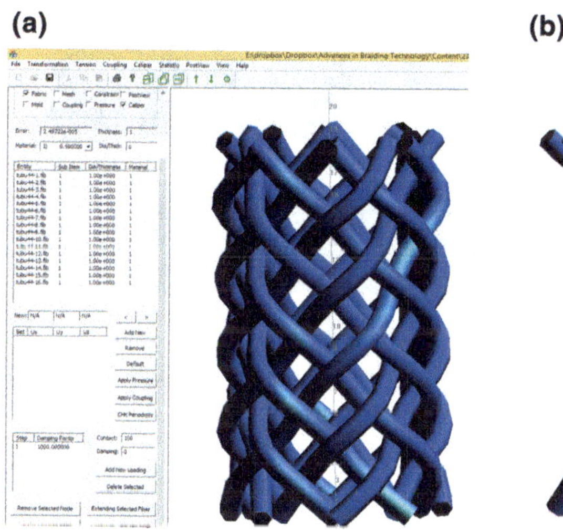

Fig. 13.7 Topological model of braid, imported in VTMS **a** as initially open structure and **b** after relaxation with VTMS [20]

with the Texmind Braider following the method of Chap. 2. After relaxation (shrinking with contact detection) the initially open structure takes very realistic geometry as presented in Fig. 13.7b. A sequence of states of another braid during axial deformation is presented in Fig. 13.8.

A simulation of the same braid after bending is visualized in Fig. 13.9.

Fig. 13.8 Braid with two yarns in a group, created topologically (TexMind Braider) and imported in VTMS for tensile test simulation

Figure 13.10 presents an application of the VMTS for relaxation of a warp knitted structure which is modelled geometrically. The examples demonstrate that the 3D geometry of the textile structures generated topologically can be simulated closer to the real structure by using the digital chain method. This is also used for various numerical analysis and optimizations.

13.5 Homogenisation for Composites

The properties of textile reinforced composites depends on the geometry, the properties of the textile structure and the properties of the matrix. The created geometrical models of the structures can be used for predicting the elasticity of the composites using different homogenization procedures. Figure 13.11 presents results of such an investigation. The geometry of a braided glass fiber beam is geometrically created

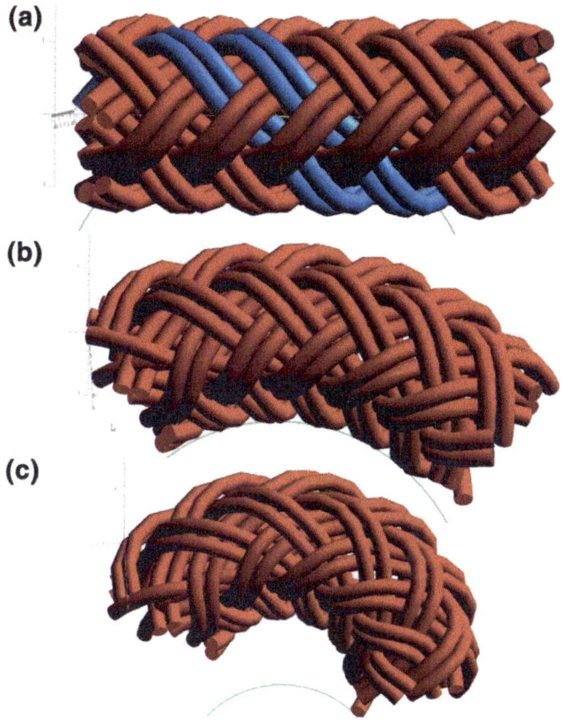

Fig. 13.9 Virtual bending test of braid, generated geometrically and imported into VTMS

and then exported to the Voxel based software, developed by G. Hello et.al., using Abaqus as a solver [15]. This way of investigation allows a quick check regarding the influence of textile reinforcement properties on the properties of the composites.

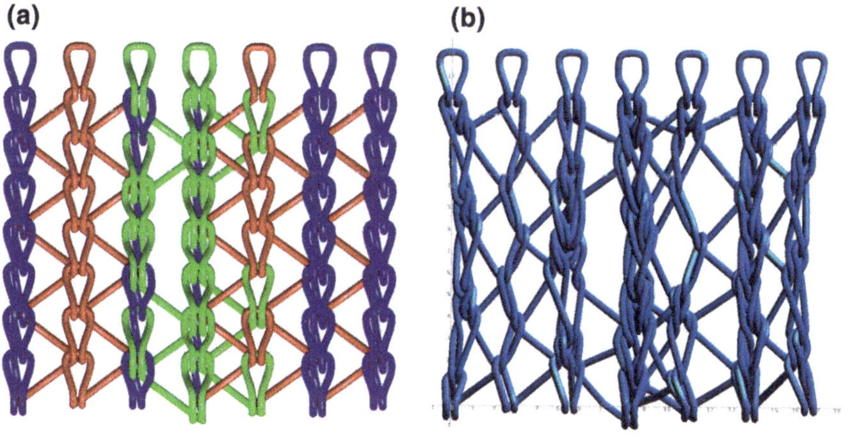

Fig. 13.10 Warp knitted structure, created using geometrical models (**a**) and (**b**) relaxed using VMTS [16]

Fig. 13.11 Braided structures with different braiding angles after voxelization and model of the complete composite beam [15]

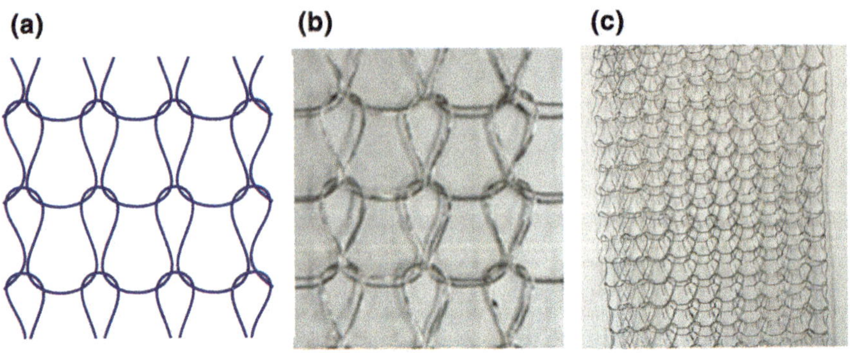

Fig. 13.12 Model and images of weft knitted hose. **a** 3D model, **b** real structure in similar magnification, **c** complete fabrics [7]

Another method for predicting the properties of textile reinforced composites it the method of inclusions [6], which is implemented into TexComp software, part of the WiseTex Suite. Wisetex already supports the XML based formats [17], which allows the implementation of exchange of data with the TexMind Viewer. This way was used for weft knitted metal wire structure Fig. 13.12. It was generated topologically using TexMind Weft Knit, then imported in Wisetex (Fig. 13.13). Once saved in the system of Wisetex the structure can be used for the creation of multiple layers with the

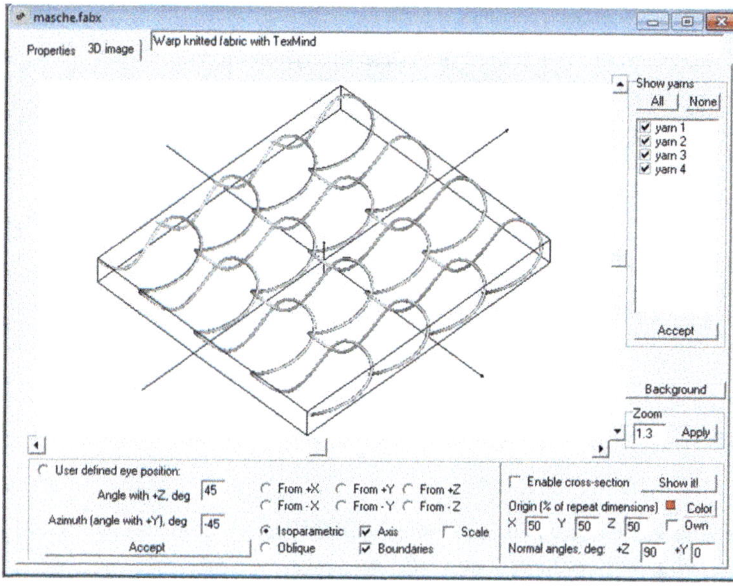

Fig. 13.13 Imported weft knitted structure in Wisetex [7]

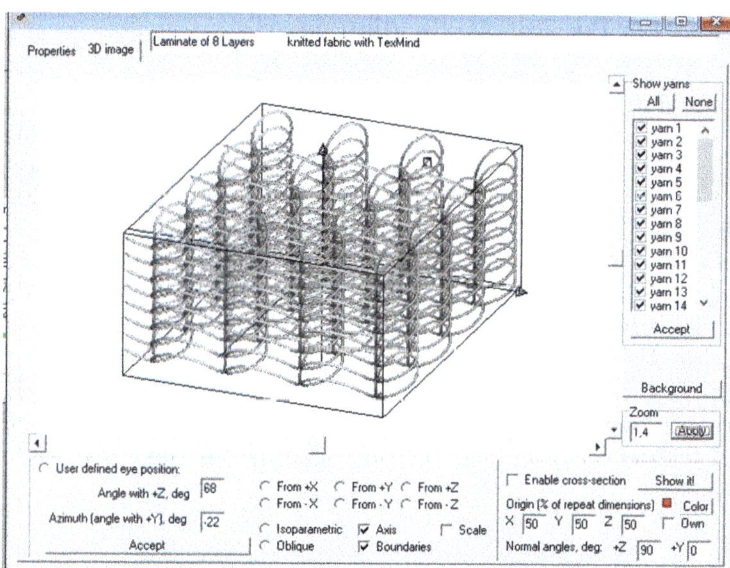

Fig. 13.14 Build 8 Layers of weft knitted metal wire structure within Lamtex [7]

LamTex program (Fig. 13.14) and then used for homogenization with the TexComp package applying the method of inclusions. The simulated wire reinforced plastic (wire-polyamid) with a wire volume fraction of 35% shows a very low anisotropy

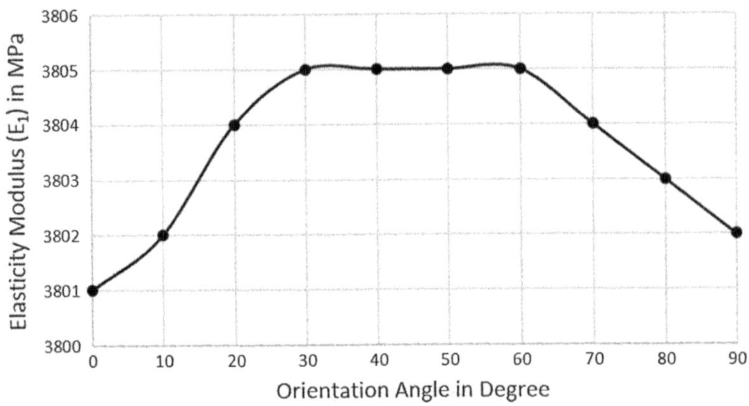

Fig. 13.15 Elasticity modulus as function of the angle, calculated using TexComp [7]

with varying values between 3.801 and 3.805 MPa, that could be negligible and an isotropic behaviour could be assumed (Fig. 13.15).

13.6 Further Possible Application Areas

The generated structures can be used as initial state for various other investigations, which are not reported here. The computation of the interaction between fluids and textile structures and assembly requires the geometry and especially the surface of the yarns or fibers. The modern CFD (computational fluid dynamics) software is able to import geometries in various formats, so investigations about the permeability and the motion of any fluid like air, water, blood or others through or around the structure can be simulated numerically. Structural-acoustic optimization [18] is another area, in which the textile structures especially for architecture and composites have to be modeled. Applying stochastic finite elements allows the consideration of the variations in properties in the textile materials and their composites [19]. Thermal properties and processes can be simulated with the finite element method too [1] and the rapid generation of the geometry speeds up the preparation of the data for such computations.

13.7 Conclusions

This chapter presented numerical simulations of textile structures, generated topologically with the algorithms of the previous chapters. The possibility for rapid generation of the yarn path data and export procedures which are especially developed for the yarns and filaments allow a significantly reduced time for the preparation of the models for simulation. For the cases, in which the simulation type is well defined, as for instance tensile test of ropes, the export procedure can write the complete solver file, so that the user does not have to prepare anything. The cases with trusses, beams, solid elements and homogenization demonstrate that the geometry, generated by using the described topological methods can be used effectively for the computation of textiles and composites.

References

1. Barauskas, R., Sankauskaite, A., Abraitiene, A.: Investigation of the thermal properties of spacer fabrics with bio-ceramic additives using the finite element model and experiment. Text. Res. J. **58**, 004051751667,722 (2017). https://doi.org/10.1177/0040517516677228
2. Cherif, C.: Drapierbarkeitssimulation von Verstärkungstextilien für den Einsatz in Faserverbundkunststoffen mit der Finite-Elemente-Methode, als ms. gedr. edn. Shaker, Aachen (1998). http://www.worldcat.org/oclc/722528088
3. Forssell, J., Mikhaylovskiy, Y., et al.: Impact (2015). http://www.impact-fem.org/
4. Freier, K.S.: Integrative produktentwicklung ballistischer körperschutz-gestricke unter anwendung der finite elemente methode. M.Sc., Hochschule Niederrhein, Fachbereich Textil- und Bekleidungstechnik, Mönchengladbach (2012)
5. Helbig, F.U.: Druckelastische 3D-Gewirke: Gestaltungsmerkmale und mechanische Eigenschaften druckelastischer Abstandsgewirke. Südwestdeutscher Verlag für Hochschulschriften (2011). http://www.qucosa.de/fileadmin/data/qucosa/documents/5207/data/Diss.pdf
6. Huysmans, G., Verpoest, I., Van Houtte, P.: A poly-inclusion approach for the elastic modelling of knitted fabric composites. Acta Mater. **46**(9), 3003–3013 (1998)
7. Kaufmann, J., Kyosev, Y., Rabe, H., Gustke, K., Cebulla, H.: Numerical prediction of the elastic properties of weft knitted reinforced composites made of plastically deformed metal wire. In: Boussu, F., Chen, X. (eds.) Proceedings of the 7th World Conference 3D Fabrics and Their Applications, pp. 217–224 (2016)
8. Kyosev, Y.: Computational model of loops of a weft knitted fabric. In: Dragčević, Z. (ed.) Magic World of Textiles, Zagreb (2006)
9. Kyosev, Y.: Geometrical modelling and computational mechanics tools for braided structures. In: Kyosev, Y. (ed.) Advances in Braiding Technology, Woodhead Publishing in Textiles, vol. 177. Woodhead Publishing (2016)
10. Kyosev, Y., Lomov, S., Küster, K.: The numerical prediction of the tensile behaviour of multilayer woven tapes made by multifilament yarns. In: Kyosev, Y. (ed.) Recent Developments in Braiding and Narrow Weaving, pp. 69–80. Springer International Publishing, Cham (2016). https://doi.org/10.1007/978-3-319-29932-7_7
11. Kyosev, Y., Renkens, W.: Numercial simulation of the tension and compression behaviour of warp kintted reinforcements. In: Binetruy, C. (ed.) Recent Advances in Textile Composites, pp. 397–404. DEStech Pub, Lancaster, PA (2010)
12. Kyosev, Y.K.: Texmind Braider (2012). www.texmind.com
13. Kyosev, Y.K.: Computational mechanics of braided structures—some case studies. In: Llorca, J. (ed.) Book of Abstracts (2015)
14. Kyosev, Y.K.: Generalized geometric modelling of tubular and flat braided structures with arbitrary floating length and multiple filaments. Text. Res. J. **86**(12), 1270–1279 (2015)
15. Kyosev, Y.K., Hello, G., Reinsch J., Aboura, Z., Schneider, J.: Determination of the mechanical properties of braided composite beams with experimental and numerical methods. In: Bogdanovich, A.E. (ed.) Proceedings (2015)

16. Kyosev, Y.K., Renkens, W.: About the 3D modelling of jacquard warp knitted structures. J. Fash. Technol. Text. Eng. S001 (2015)
17. Lomov, S.V., Verpoest, I., Cichosz, J., Hahn, C., Ivanov, D.S., Verleye, B.: Meso-level textile composites simulations: open data exchange and scripting. J. Compos. Mater. **48**(5), 621–637 (2014). https://doi.org/10.1177/0021998313476327
18. Marburg, S.: Numerical damping in the acoustic boundary element method. Acta Acust. United Acust. **102**(3), 415–418 (2016). https://doi.org/10.3813/AAA.918958
19. Sepahvand, K.: Stochastic finite element method for random harmonic analysis of composite plates with uncertain modal damping parameters. J. Sound Vib. **400**, 1–12 (2017). https://doi.org/10.1016/j.jsv.2017.04.025
20. Zhou, E., Mollenhauer, D., Larve, E.: A realistic 3D textile geometric modeling. In: ICCM-17: 17th International Conference on Composite Materials. IOM Communications, London (2009)
21. Zhou, E., Sun, X., Wang, Y.: Multi-chain digital element analysis in textile mechanics. Compos. Sci. Technol **64**(2), 239–244 (2004)

Summary About the Topology Based Methods

This book summarizes the textile specific rules for the modelling of interlacement of yarns in knitted and braided structures. The implementation of these rules in algorithms and software allows textile engineers to visualize in 3D, check and optimize these structures in the same way as it is already possible for woven structures and for parts and assemblies of metal or other homogenous materials which is already possible for a long time. The rules about the interlacing are described in a "topological" way. It is done with parametric definitions of the coordinates of important key points of the yarns which are related to the other yarns. For regular (tubular and flat) braids this method is better than the methods reported in literature, because it is generalized. This means that they are applicable for *any* floating length and *any* number of yarns in a group. The analysis of the structures and the numerical results lead to several essential details in the algorithms that need to be taken into account in order to keep the results correct and close to the reality. Contrary to the weaving topology, the braids can only have a limited number of inlays between the braiding yarns, depending on the number of horn gears and not on the number of crossing yarns. Even though braids have unit cells that are identical to woven structures, only a very small number of patterns produced in weaving can be produced in braiding. These two aspects have usually been skipped by most of the researchers that have covered this topic previously. The model presented in Chap. 2 is based on the production process of the braids and overcomes these disadvantages. It remains a topological model with all its problems about the intersection of the yarn volumes at some angles, but it covers all kind of tubular and flat braid configurations. Chapter 3 demonstrates that the created geometries are useful in the engineering design of these braids and can give approximate evaluations of the jamming states, diameter—braiding angle relation, weight per meter, required yarn length and characteristic points of the force-elongation diagram.

The topology of more complex braided structures can be generated in the same way as of regular structures, but require one additional level of generalization. The arrangement of the horn gears for regular braids (tubular and flat) follows some rules which are considered in the development of the models for these structures in Chap. 2.

Y. Kyosev, *Topology-Based Modeling of Textile Structures and Their Joint Assemblies*, https://doi.org/10.1007/978-3-030-02541-0

The correct configuration of the horn gears for any **custom** braiding machine is only known by its developer or user of the ready machine, but not by the model developer. In this case, for creating the topological model of the braids the machine configuration has to be reconstructed virtually and the carrier motion has to be emulated completely. Two efficient approaches—horn gear based and track based—for such models are developed, implemented, evaluated and described in Chap. 4. The efficiency of the approaches is connected with their limitations, therefore, the future clearly lies in a combined approach in which the tracks and the gears are modelled completely. Still, both of the approaches allow the modelling of custom braiding machines with fixed tracks. They present more than 99% of the braiding machines used now and their structures can be generated by using the topological methods developed and presented in this book.

Knitted structures are based on completely different types of interlacement (interlooping) than braided or woven structures. The interlooping within knitted structures is based on a three dimensional orientation of the yarns. Several researchers tried to reduce the dimensionality of the knitted structures using the single crossing places as cells, but this approach was only applied successfully for basic types of structures. The powerful and generally applicable way for modelling knitted structures requires structural elements, which are connected (again, like in the braiding) with the production process. Actually, both in the weft and warp knitting the parameters of the production process influence the appearance and the *geometry* of the structural elements. The available yarn length for (multiple) tucks in weft knitting determines whether a nice small opening with visible several yarn pieces will be built or if a three dimensional element with increased thickness like a nope will be produced. Topologically, both cases are the same but geometrically and finally mechanically the yarns have different configurations. The situation is more complex for platting loops and especially for platting loops in warp knitting. The position of the undermapping is well defined—this depends on the guide bar arrangement only. The position of the loop heads and legs can vary, depending on the pattern, yarn tension and guide bar position. In some cases, the one yarn in plating loop becomes visible in other case the another one. Topologically, at the scale of complete cells, both configurations are strictly seen as identical, but for the local regions these are different configurations. Chapter 5 describes these issues with practical examples from the modelling experience of the author and finally demonstrates that in any case the topological way for initial creation of the geometry is the *fastest* way of geometry creation. The results of such models are implemented in two commercial packages. Both are used in the industry which is a proof of the usability of the results. One way to improve the appearance of the topologically generated knitted structures based on truss framework model is described in Chap. 6. The model considers the orientation of the yarn pieces and uses the loop centers as joints. It works perfect and simple for several classical warp knitted structures, but as it is demonstrated in the chapter, its solvability depends on the boundary conditions of the unit cell. The linear system of equations is sensible and can under-determinated or over-determinated with small change in the pattern or the type of the boundary conditions. The model itself only works for loop based structures and only some structure with thin weft yarns, because the thick or multiple weft yarns stabilize the loops. Last but not least, it is only applicable for small deformations which are not including friction and sliding. Due to these reasons, this model is not suitable for general cases of refinements, but has its advantages and can be used for draping simulations or other situations in which the loop centers remain fixed.

Chapter 7 only states "*Notes* about topological methods for woven structures" and presents few basic sources, approaches and problems in the modelling of woven structures. The woven structures are not covered within this book, as their models are developed significantly better at geometrical and mechanical level. The aim of this book is increase the level of topological modelling of braided and knitted structures up to a level at which the modelling of the woven structures is today. As mentioned before in the chapter, the challenge of the topology of woven structures is returning with the introduction of the latest development of weaving machines. Computer controlled frames, weft insertion devices and open reed techniques allowing the stitching of similar structures are some examples. All these machine developments will require suitable advanced developments of the way of "topology" coding of woven structures.

Any of the textile structures is usually sewed together to assemblies. The kind of stitch formation used and its influence on the properties of the final product are important. Modern simulation software, usually based on FEM, is able to simulate the behaviour of textile yarns, but requires their complex geometry. This topic is covered in Chap. 8 which demonstrates the topology based approach for a few basic stitches. How to orient the simulated fabrics in the space and prepare for stitching—is mentioned in the Chap. 10. The methods are known in the computer graphics, but there are several implementation issues for the yarn based level which has to be considered in order for the results to be useful for the textile people. As for instance—how to model the cutting of the fabric in order to preserve the topological information at yarn level.

All models are based on yarn path description, because the *yarns* are defining the topology of the textile structures. Because the yarns are made from fibers, they are deformable. One promising way of modelling the deformable yarns is the use of the filament or fiber level models. Promising, because such modelling eliminates the need of definition of *continuum*-similar properties of the yarns in lateral direction, and concentrate in the natural way of modelling of fiber based structures—modelling the fibers or filaments. How to fill the generated paths with filaments for further calculations at that level is explained in Chap. 9. This chapter also contains some visualized comments on the curvature of the path curve and its influence on the real and the artificial twist of the yarns.

After all yarn based structures and their connecting stitches are described, the book finally demonstrates examples of the use of mechanical models for computing properties or at least refinements of the topologically generated structures. Chapter 12 reports about the author's own implementations for the integration of the motion equation of the textile structure, based on mass-spring model as discretisation of the partial differential equation for the motion of the one dimensional continuum. The Verlet algorithms with some damping allow simulations of the unwinding process and of axial loading of weft knitted structure. The results show that the method works, but requires efficient implementation of the contact search strategies and algorithms in order to be applicable for industrial use. Chapter 13 demonstrates the another approach—the topologically generated geometry is imported in suitable format in external FEM, digital chain or homogenisation software and the calculations are performed with these. Applications for the optimization of structures for ballistic applications, calculations of the mechanical properties of composite structure, or deformation of braided structures are demonstrated.

All algorithms described in this book are implemented in a suite of software packages, successfully used for research and education in the worldwide industry. The comparison to another similar package for the 3D visualisation of textiles in Chap. 11 demonstrates that the developed algorithms and software fill gaps which were not at all or only partially presented until now. Opening the possibilities for 3D visualizations of the textile structures is an important contribution of this book. The visualization helps a lot in teaching and for the first evaluation of the suitability and the correctness of a structure. The application of the generated models for direct evaluation of the properties of products (such as cover factor, specific mass) is another aspect, where the developed methods helps and which is valued by the industrial people. The possibilities to export the generated structures and then to refine these with external tools and to perform complex mechanical simulations is as another contribution which is highly appreciated by regrettably only few well qualified researchers. Regrettably only few, because the people who are able to perform correct mechanical analysis with friction, nonlinear material and geometry with modern FEM software are rare and the majority of these people do a good job in industrial conditions and have no rights to show these to the publicity.

The book is aiming to fulfil two tasks. The first task is to increase the number of people understanding the problems of textile structures and their modelling by presenting the author's experience. The second task is to give an understanding of the topological methods and their problems. This shall help to create more powerful models so that the industry can create better products and can solve it although "we, the model developers, are only few".

Outlook

One of the areas of future works is already mentioned in the previous section and consists of extension of the models in order to become more suitable for computer controlled braiding and weaving machines. The computer controlled machines make the interlacement positions dynamical and the generation of the topologically, geometrically or mechanically based yarn architectures can be done by using only numerical methods with time integration. Several of such tasks in braiding and weaving are already in progress, but are not finished. Thus, they were not included in this work. Braiding machines with computer controlled switches and horn gears are installed in more and more laboratories. This leads to an increased need of user friendly tools for the design of products made by these machines and therefore force the development of models in this direction. Software and models for square sets of gears for the two large 3D-braiding machines exist since several years, but they are not extend-able to configurations with different sizes of gears for which the square set model does not work. The algorithms presented in the chapter for braiding machine emulation with extension for time controlled dynamical switches will allow the creation of CAD tool for these modern machines.

Generally, the topological models for all kinds of structures were created as an open structure with larger distances between the yarn in order to avoid yarn intersection during the visualization. This approach is required for the generation of clear and topologically correct textile architectures. It requires a relaxing step after the geometry creation, in which the yarns shrink and the structure becomes compact. Or, if the structure is generated with thinner yarns than in reality, a step of (thermal) expansion is required. Both techniques are efficient and applied by various researcher, but they present an artificial way for refinement of the topologically generated structures. The natural way of generating the virtual structure is and remains the complete simulation of the textile *process*. It only uses the process simulation and all its factors, like different tensions, irregularities on the machine etc. can be considered and their effects can be effectively seen in the generated structure. Therefore, in the long time scale, the topological models will remain tools for rapid *pre*-evaluation of specific products, before the more time consuming textile process simulations are started. Still, the implementation of the process simulation at applicable level for the non-FEM and

Y. Kyosev, *Topology-Based Modeling of Textile Structures and Their Joint Assemblies*, https://doi.org/10.1007/978-3-030-02541-0

non-programming educated people will take a few more years. The principle algorithms are existing already and have proven that they work, but the computational time and the requirement of a highly qualified user is unacceptable for industrial application, even today. In order to make these algorithms and methods applicable, a lot of work in the improvement of their efficiency and stability is required. This work concern mathematicians, programmers and mechanical engineers. It requires the development of well *parallelized* algorithms for computation. Such are existing for linear system and differential equations, but as the book demonstrates, the textile structures have a lot of contact places. This means that the data between their participants has to be exchanged. Data exchange means the reduction of the efficiency of the computing. Finding a suitable way for the parallelization with contact detection and mechanically correctness using GPGPU will significantly speed up the computations and will a new era in the simulative investigation of textile structures. Of course this will have to include multiple scales or at least yarn and filament levels. In the area of mechanical behaviour the rheological models of the polymers have to be included. The current efficient algorithms compute with pure linear spings or even with undeformable yarns. The yarns often show a time-depending behaviour, which depends additionally on the load history including damages. Integration of these effects in the models will allow a more effective and more accurate computer assisted engineering design of textile products.

MIX
Papier aus verantwortungsvollen Quellen
Paper from responsible sources
FSC® C105338

Printed by Books on Demand, Germany